# Environmental Cell Biology

John LaFollette Howland
Bowdoin College

**W.A. Benjamin, Inc.**
Menlo Park, California
Reading, Massachusetts
London • Amsterdam
Don Mills, Ontario • Sydney

*To Cynthia, Ethan, and Hannah*

Cover drawing of prairie falcons is adapted from a photograph by Robert J. Erwin and Richard D. Porter.

Copyright © 1975 by W.A. Benjamin, Inc.
Philippines Copyright 1975

All rights reserved. No part of this publication may be reproduced, stored in a retrieval system, or transmitted, in any form or by any means, electronic, mechanical, photocopying, recording, or otherwise, without the prior written permission of the publisher.
Printed in the United States of America.
Published simultaneously in Canada.
Library of Congress Card Catalog Number 75-1992

ISBN 0-805-34520-5
ABCDEFGHIJKL-AL-798765

*W.A. Benjamin, Inc.*
*2727 Sand Hill Road*
*Menlo Park, California 94025*

# Preface

Public interest in ecology and especially in the impact of humanity upon the environment has become intense in recent years and pollution is being examined sternly by the press and by scientists of many disciplines. It appears to us that relatively little attention has been given to the effects of pollutants upon the processes occurring in cells and that such information should nonetheless be helpful in judging the various components released into the environment. While toxicologists have, in fact, been engaged in studying such questions, very few of their results have reached publications intended for the general audience or for students at the introductory level of science. It is to this last audience that *Environmental Cell Biology* is directed. Rather than providing a complete summary of all that is known about cellular effects of pollution, this book represents a selective outline which defines the field of pollution physiology and biochemistry and may suggest to the reader strategies for pursuing it.

Since this book is an introduction and since it is important to motivate students in a variety of scientific areas to be concerned with this topic at an early stage in their education, it will be assumed that the reader has had minimal course work in either chemistry or biology. More specifically, we assume a level of exposure to both chemistry and biology achieved in secondary school. For this reason, and because the nature of our topic requires that we discuss various organic chemicals such as pesticides and a variety of physiological processes such as respiration, we will proceed slowly and introduce as much cell biology or chemistry as needed en route. To this end, Part I (Chapters 2-4) introduces the basic concepts of cellular structure, cellular response, and ecological systems. Thus the reader is prepared for the specific cases presented in Part II (Chapters 5-8) including cellular effects of insecticides, atmospheric pollution, metals, and carcinogens.

The nature of this topic also requires that certain organic chemical structures are set forth throughout the text. If the reader has taken a course in chemistry, he may only wish to look at the structure, see what it tells him, and move on. However, the less chemically oriented reader may find it advantageous to turn to Appendix I which provides basic chemical formulas. The reader should be assured that detailed knowledge of organic chemistry is not required to understand the main features of the subject matter of this book.

In writing *Environmental Cell Biology,* the author was aided by the suggestions of Professor Charles E. Huntington and of Professor C. Thomas Settlemire, who also contributed the chapter on metal pollution. During the writing of the book, the author was recipient of a U.S. Public Health Service Research Career Development award. The author also wishes to thank Mrs. Virginia Richardson, Mrs. Nancy Payne, and Mrs. Barbara Hess for their valued assistance in typing portions of the manuscript.

Harpswell, Maine                                                           John LaFollette Howland

# Contents

Chapter 1    Introduction    1

**PART I**    **Basic Concepts of Cellular Biology**    5

Chapter 2    Ecology of Polluting Substances    6

Chapter 3    Cells and Basic Physiological Systems    19

Chapter 4    Cellular Protection Against Pollution    51

**PART II**    **Cellular Aspects of Pollution**    63

Chapter 5    Physiological Effects of Insecticides    64

Chapter 6    Atmospheric Pollution    84

Chapter 7    Metal Pollution    95

Chapter 8    Additional Forms of Pollution    104

Chapter 9    Conclusions and Suggestions for Further Reading    112

Appendix: Basic Organic Structures    116

Index    121

# Chapter 1

## Introduction

**The Scope of the Problem**

Anyone who observes the effluent pipe from a paper mill will not require a very academic definition of the word *pollution*. The common view of pollution is that it is material deposited in the environment by humans and that it is either harmful to health or to the esthetic sense. This definition is perfectly adequate in such situations. Obviously, the release of chemicals which are harmful to health and wellbeing does not constitute existing in harmony with the environment, and it is equally clear that such activities should be minimized.

Although pollution is not a new problem, it received a sharp stimulus by the new capabilities developed during the Industrial Revolution. The chemical industry that grew during the past century has led to the production and dumping of compounds of great variety and, in many cases, great toxicity. In a number of instances, toxic compounds were developed *in order* to be dumped into the environment, for example, to kill insects, and in many cases they were specifically devised by synthetic chemists to be as little degradable as possible. Such persistent insecticides were desirable because they remained active for a long time. Indeed, the turning point in concern about pollution could be defined as the time when persistence in insecticides changed from being a virtue to being a grievous shortcoming. This turning point came a remarkably short time ago. In fact, serious opposition to indiscriminate pollution has existed for little more than a decade with the publication of Rachel Carson's *Silent Spring* in 1962 being widely regarded as a milestone. Opposition that is informed about the harm caused by pollution is still in the process of developing, and the sort of information that could lead to major reduction of some kinds of pollution, for example, by providing less harmful alternatives, is fragmentary. Some of the complexities involved in attempting to reduce pollution are mentioned in later chapters.

Most insecticides are novel compounds in the environment having been created for the task by the inventiveness of the chemical industry. A few of them are natural products such as nicotine and the pyrethrins, which produce smaller dislocations in an environment that is, to a degree, adapted to their existence. Although much of the treatment of this book necessarily regards insecticides as harmful compounds to organisms besides the target organisms, insecticides have played a major role in increasing crops and combating disease. This role has been specially significant in the Third World—the developing nations. Most people who regard insecticides as universally vile poisons are not inhabitants of the Third World and are most likely rather uninformed about the concerns that occur there. If one condemns the use of insecticides out of a desire to exist in harmony with the environment, then it is important to be in harmony with more of an environment than that found in suburbia alone. In other words, the pollution problem cannot be considered in any context less than the whole world.

When beginning to consider the cellular influence of substances associated with pollution, it is best to avoid emotional interpretations. There is, for example, a tendency among those who are concerned but marginally informed about pollution to regard material in pollution as intrinsically harmful, while the same material if it occurs naturally is regarded as benign or not thought about at all. A good example is the furor existing for several years about mercury levels in the sea and in food species of fish. The reaction to the finding of mercury in fish has included governmental action and much private anxiety. The discovery that preserved specimens of fish which were caught before much mercury pollution occurred exhibit practically the same levels of the metal did not affect the harshness of the reaction. From many of the statements made at the time, it became evident that normal environmental levels of mercury were acceptable to many people while the same levels, if they resulted from pollution, were not. Obviously, when cells are influenced by mercury they do not know where the mercury came from. A similar argument applies to other forms of pollution and to the notion that cells know the history of their nutrients. Some people hold the interesting view that a vitamin is beneficial if extracted from a carrot but useless if produced in a factory. A molecule is a molecule, and the notion that they carry with them a record of their history seems to be a particularly primitive one.

## Some Concepts and Definitions

Before proceeding to physiological aspects of pollution, we first consider several additional general features of the problem. We have already mentioned the burden of emotion that often interferes with objective thinking about environmental matters. To avoid this obstacle, let us reflect on some concepts and definitions which should promote clarity.

First, our present study is *environmental toxicology,* which is the study of undesirable effects of environmental conditions and substances upon living organisms. Since we will be chiefly interested in cellular effects, our endeavor might be called "cellular environmental toxicology." However, that title is long and pompous and we shall not use it. The point of our above definition resides in the word "undesirable." One might ask how the criterion of desirability (or undesirability) is to be applied. In other words, one is permitted to ask "undesirable to whom?" which leads to additional questions of some profundity. Thus, an herbicide is obviously undesirable from the viewpoint of the plant and an insecticide from that of the insect. Of course we tend to regard effects on humans as the touchstone of undesirability, and that is a proper approach as long as we are considering only direct effects such as brain damage resulting from ingestion of lead. However, it does seem arrogant to suppose that toxicity to humans should be the only test; it is also probably poor ecology as seen in Chapter 2. If exterminating mosquitoes upsets a larger ecological pattern, for instance, by depriving certain insectivorous birds of their food supply, then there may be perilous results if we exterminate without prior careful investigation.

Additional technical and philosophical problems are associated with determining the adversity of an effect. For instance, we are far from a complete understanding of the normal condition with regard to many human physiological processes. Moreover, humans exhibit a wide variety of physiological attributes, and a wide spectrum of values may be observed with any physiological measurement selected as important. Thus, it is often very difficult to judge whether any adverse effect has occurred in a particular human; we must have recourse to statistical treatment of such measurements, and statistics are often subtle and sometimes misleading. In any event, in the final analysis it is the individual human who feels the harmful effects if the environment becomes toxic, as the weighted average of a physiological value "x" does not have any capacity to feel pain or outrage. Furthermore, we can never be certain about the safety of a particular compound since there is always the possibility that a new physiological test will demonstrate a harmful effect (just as we can never absolutely prove the truth of a theory because a future experiment might invalidate it).

While one can appear to avoid the problem of choice of physiological measurements by considering more general tests of health, such as life-span or mortality, this strategy leads to other problems. For example, many toxic agents such as heavy metals require years to exercise their harmful effects. Others such as those causing mutations, or alterations, in cells might only influence the subsequent generation. Some agents might produce effects so variable that the only way to be certain about details of their toxicity for people would be to conduct controlled experiments with humans which is, of course, highly unethical and thus impossible. Finally, even if we are able

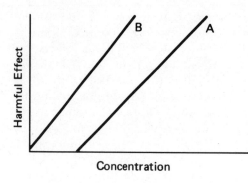

**Figure 1-1.** Harmful influence of polluting material with a threshold (A) and without a threshold (B).

to prove that a given component of pollution has no effect on the human life-span, how are we to assess its influence on the quality of that life? Subtleties abound and easy answers are frequently wrong.

A public desire to identify its enemies in the case of pollution is understandable but sometimes not very helpful. Thus, zinc is often regarded as "bad" when it is dumped into rivers in large amounts, but it is also a normal component of living organisms and some of the enzymes that support the life of humans exhibit an absolute requirement for the zinc ion. One is forced to conclude that "the enemy is within us," and the black-and-white determination of friend and foe is in shambles.

Finally, even if we temporarily ignore the problem of components of pollution that are, in smaller amounts, nutritional requirements, we must still consider a related problem, namely the determination of whether there is a safe level of exposure to *any* components. Thus, we might ask if there is a "safe" concentration of mercury in seawater. This is equivalent to asking if there is a "threshold" concentration below which no harm can result. The two possibilities are illustrated in Figure 1-1; either there is a safe level in the case of A or there is not as in B. Since the environment in its unpolluted state may contain some of the substance anyway, this sort of question is always difficult to answer. If one is to be practical he should consider that we all carry the burden of proof to show that there is a completely safe level. It is always preferable to err on the side of care and to work to keep known harmful agents at the lowest possible concentration in order to at least minimize harm, if not to eliminate it.

# Part I

# Basic Concepts of Cellular Biology

# Chapter 2

# Ecology of Polluting Substances

Ecology is the science that deals with the miraculous interrelations among organisms and interactions between organisms and environment. It is also a word that has been given a rich overlay of meaning by some of the public, denoting everything from living in harmony with the universe to refusing to throw beer cans out of car windows (the two not necessarily being unrelated). Moreover, the science of ecology has been appointed by the popular press as the saviour of mankind, apparently replacing the Industrial Revolution, which has been relegated to a considerably inferior role.

Since in this new theology pesticides and other polluting materials are at the basis of much evil and since names like *dieldrin* and *heptachlor* are widely known, we will begin our study by placing pollution into a more ecological framework. For one interested in physiological aspects of pollution, there are also some practical benefits to be gained from such a consideration. For instance, one can address the problem of the manner in which polluting chemicals move through the natural world. One can also see in ecological perturbations the final results of pollution in the sense that ecology reflects the sum of the physiology of the members of a population. In other words, if a physiological influence of a substance is profound enough, it is likely to produce an effect on populations, and this effect will undoubtedly influence other populations as well, cascading through the *ecosystem*, or ecological system, in a complex and often dramatic fashion.

## Ecosystems and Populations

An ecosystem may be roughly defined as the portion of the world with which one is concerned in a given ecological problem. The ecosystem *is* the world when the global nature of a problem demands such a scale, as in the case of the worldwide distribution of DDT. In other instances it may be a few cubic millimeters of soil, as when one studies soil bacteria. In any case, the point is that one must decide what boundaries are reasonable for a particular problem, and then one must be very sensible and flexible about enlarging them if necessary.

A more precise definition of an ecosystem requires consideration of both the nature of the boundaries that enclose it and the components of which it is composed. As just stated, boundaries must be relevant to the particular problem under consideration. This means that none of the important ecological transactions occurring in the system should occur across the boundary. In other words, if predator eats prey or if the prey has just fed on a plant or if the plant has absorbed potassium from the soil, all components, living and nonliving, should be part of the ecosystem. Similarly, when a food chain or a biological cycle of matter occurs in the ecosystem, boundaries should ideally be set so that it occurs entirely within. In most instances it is desirable to regard an ecosystem as "closed" with respect to such transformations of material as represented by food chains or biological cycles. At the same time, the ecosystem is "open" with respect to energy in that the primary source of energy is the sun, which is external to all ecosystems, and also in the sense that energy is eventually lost from the ecosystem largely through radiation of heat. Some features of the boundaries of ecosystems are analogous to those of "isolated" systems important in thermal physics; in both cases the degree of isolation is invariably less than complete in any instance but a hypothetical situation, while in both cases the fiction of isolation greatly aids in considering the structure of the system.

Since much of the business of ecology is analysis of the structure of ecosystems, it is important to be familiar with some of their components. First, an ecosystem consists of a portion that is nonliving, or *abiotic,* and one that is alive, or *biotic.* The nonliving part is sometimes considered the environment, but this is certainly misleading as an organism's environment includes other living organisms, too. The nonliving part of the system is largely inorganic, for example, the water of the sea, nitrogen gas of air, or the minerals of soil. But it contains organic matter as well, largely in the form of residual material from dead organisms. The living part of the ecosystem exhibits enormous variety that can, from an ecological viewpoint, be organized according to food sources of the different organisms.

After defining an ecosystem, it is necessary to decide what features within the boundaries need to be monitored or discussed. One must consider what organisms constitute the system, what interrelations between the organisms, such as predation, are of interest in the context of the study, and in what manner energy and nutrients flow through the system. In the context of this chapter, one must also consider the process in which pollution enters the system and how it flows through the various populations within it. In many cases, a study of the effects of pollution on populations can be very instructive in reaching an understanding about the relationships between populations themselves. Thus, pollution can be regarded as a huge, grotesque experiment which was carried out quite unintentionally but an experiment from which a great amount may be learned.

## Food Chains

Polluting materials do not affect all organisms in an ecosystem equally, and they pass through the system from one class of organism to another by routes that are often predictable. In fact, the way such substances move in the ecosystem is analogous (and sometimes identical) to the movement of nutrients and energy through the same system. For this reason we first discuss the idea of a *food chain*, which may be roughly defined as the order in which organisms eat or are eaten. Size is a major determinant of the direction taken by the chain. Thus, it is not surprising that the blue whale eats krill (a small shrimplike organism) and not vice versa, although examples of smaller animals eating larger could also be cited. In fact, food chains should really be thought of as circular, at least in a well-regulated ecosystem. Thus, the final predator, such as the ecologist himself, is finally allowed to return to the nutrient pool owing to the efforts of the "conqueror worm" (as well as conqueror bacteria, moulds, etc.).

Food chains are not completely cyclic inasmuch as they all begin at the point where sun falls upon a green plant, allowing photosynthesis to occur. This is the point at which most energy enters the ecosystem from without, the energy being produced in thermonuclear reactions in the sun. Energy from solar radiation is used to produce ATP (the high-energy compound of cells) and to carry on certain other chemical reactions. One of the net results of photosynthesis is the reduction of $CO_2$ to form reduced carbon compounds from which cells are made. This process is called *primary production* and is often measured by allowing a plant (or portion of the ecosystem such as the plants in a volume of seawater) to photosynthesize in the presence of radioactive $CO_2$ where some of the usual ($^{12}C$) carbon atoms are replaced with radioactive $^{14}C$. One then measures the amount of radioactivity incorporated into cellular material and can calculate the amount of living material (*biomass*) produced.

Green plants are termed *autotrophic* since they can grow without using other organisms as sources of organic materials. In the same category are the lower photosynthetic plants, including algae and photosynthetic bacteria. These organisms are consumed by animals called the *primary consumers,* which are herbivores. This group includes the zooplankton that graze on the autotrophic phytoplankton living in surface waters. The primary consumers, in turn, serve as food for *secondary consumers*—the carnivores. Then there can be several steps in the carnivorous ("dog-eat-dog") part of the chain, terminating in the final consumer of which the reader of this book may be a good example.

If one conducts a primary production experiment with radioactive $CO_2$, one sees that the algae and other green plants are the first organisms to be labeled with $^{14}C$, the primary consumers second, and so forth. Thus, the particular carbon atoms identified begin in the $CO_2$ of the atmosphere, are "fixed" into biological carbon compounds by the photosynthesizing organisms, and then flow through the ecosystem through the various levels of consumption. A carbon atom that began in the air over an ocean might spend a period of time as part of a protein molecule in the chloroplast of a kelp plant, thence to the liver of an herbivorous fish, and "temporarily" end its career in the left thumb of a human. It is thus with polluting substances as well; one can trace a similar history of such compounds, and several features are interesting.

In the first place, polluting substances must be considered in the context of their *solubility,* which is defined as the amount of the substance that can be dissolved in a given amount of another substance. For example, a substance

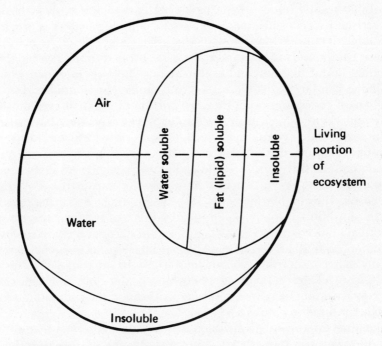

**Figure 2-1.** A hypothetical ecosystem compartmented according to solubility. The living, or biotic, portion of the ecosystem is shown as the circle-within-the-circle. It is divided into the water compartment, which is the largest part of most cells; the lipid compartment, which includes the cellular membranes; and the insoluble part which includes, for example, cell walls and mineral deposits. The insoluble part of the abiotic portion, shown on the bottom, represents sediment and ocean floor in the aquatic environment and soil on land.

that is dumped into the sea and which is highly insoluble in seawater will have difficulty entering the biological part of the sea unless there is a special mechanism promoting it. In other words, a compound that precipitates to the sea bed immediately is little threat to the biomass of the ocean.

One instructive way to consider an ecosystem is in terms of a collection of compartments based on different solubilities. This approach is diagrammed in Figure 2-1 where a hypothetical ecosystem is subdivided according to what substances might dissolve in the several compartments. Thus, a compound that is soluble in air, such as carbon monoxide, would be a particular threat to those organisms that have an extensive air interface provided, for example, by lungs. More importantly, a fat-soluble compound (one that is soluble in fats and other organic substances providing solution) that is dumped into the ocean might be expected to become concentrated in the fat-soluble compartment of the biological part of the ecosystem. This is not a trivial process since many important polluting materials are, in fact, fat-soluble. These include most of the insecticides as well as many other noxious compounds. They tend to be concentrated in the fat-soluble compartment in our diagram and especially in cellular membranes, which are about one-half fat (lipid). Thus, DDT and similar compounds end up at relatively high concentration in the very portion of the biological part of the ecosystem where, as we will see in Chapter 5, they exert their harmful effect.

Since fat-soluble compounds become concentrated in the fat compartment of organisms, this concentrating effect tends to become magnified as the compounds travel up the food chain. For instance, an insecticide may be absorbed by algae which are eaten by primary consumers which, in turn, are consumed by higher members of the chain. At each step in this process, the concentration of insecticide in the lipid part of the organisms becomes magnified. The compound may be present in water at a relatively low concentration such as 1 micromolar.[1] The primary producers that take it up initially may concentrate it about tenfold (to 10 micromolar). This is not through any effort on their part; they merely absorb it from the seawater. The concentration increase occurs simply through the compound's greater solubility in lipid as compared to seawater.

The same effect can be illustrated in a test tube. If some water and ether are added in a test tube, they will not dissolve in each other (to any great extent) but will instead form two layers. A lipid-soluble material, such as

---

1. A concentration of 1 micromolar means 1 micromole (one-millionth of a mole) dissolved in 1 liter of solvent, the substance providing solution. One mole of a substance is the molecular weight of that substance expressed in grams. Thus, if the molecular weight of DDT is 354 units, then a 1 molar solution of DDT would contain 354 grams per liter, and a 1 micromolar solution would contain 0.000354 grams per liter. This is, of course, a very dilute solution, and it is characteristic of many insecticides to be effective at such concentrations.

insecticide, can then be added to the water part. When the tube is shaken, the insecticide will be concentrated in the ether fraction, although none was there at the start of the demonstration. Organisms react similarly, and they are able to do so repeatedly as they are ingested by organisms further along the food chain. Remember that an organism feeds on the contaminated prey repeatedly during its lifetime so that its internal concentration of the compound will rise throughout its lifetime until it becomes, in turn, prey for an organism further along the chain. Thus, the compound that might be present at a level of 1 micromolar could reach 100 micromolar after a few steps in the food chain. A concentration of 1 micromolar might be quite harmless while 100 might be lethal. This simple effect accounts for considerable discussion between insecticide critics and apologists; the environmental amounts of many pollutants are truly harmless as long as they are evenly distributed, but nature doesn't leave them that way.

Such food-chain magnification of concentration accounts for the well-known fact that birds are more harmed by DDT and related compounds if they are high on the food chain. Thus, falcons are presently threatened due to the thin eggshell syndrome (discussed in Chapter 5) since they feed on other birds that are high on the food chain and have already concentrated DDT to a near harmful degree. Seed-eating birds are lower on the chain (and, if anything, serve as food for the falcons) and, significantly, have lower levels of DDT in their lipid fraction.

## Mixing and Residence Time

In considering the fate of polluting compounds that enter the environment (not to mention the fate of the environment itself), it is important to consider the mechanism of dispersal. For example, an aerial photograph of an industrial plant in the act of dumping paper waste into a river will show a plume of discolored water beginning with the effluent pipes and extending downstream for perhaps a mile before becoming increasingly indistinct. The rate at which the effluent becomes diluted (mixed with water) will depend upon such factors as the amount of effluent being dumped per minute, the flow rate of the river, and perhaps the topography of the riverbed. To assess the effect of the pollution on river life, the distance from the pipes must be specified. There will undoubtedly be an expansion of effects with, for example, nothing but bacteria growing in the initial part of the plume and essentially normal river populations occurring a few miles downstream.

Mixing is by no means the only influence that must be considered in this type of case. For instance, the bacteria growing in the river (largely as a response to the selective pressures produced by the pollution) are also engaged in breaking down many of the compounds constituting the pollution. Their effectiveness in cleaning up their environment will also depend on many of the same factors such as flow rate of the river, but also others that influence the bacterial growth rate, such as temperature.

A flowing river is obviously a better vehicle for removing and mixing a liquid pollutant than is a relatively static body of water such as a lake or ocean. The polluting material thus may be only very slightly mixed with the larger volume of water so that very high local concentrations occur where the material is injected into the body of water. Moreover, the body of water may be structured so that there is never complete mixing of the pollution into the entire volume. In a very deep lake there may be only slight communication between the surface and deep waters so that pollution is confined for a long time to the relatively small surface volume. This high concentration in surface layers is of severe consequence since the life of the entire body of water is dependent upon primary production (photosynthesis) at the surface. If photosynthesis is eliminated, all populations of the body of water are likely to be exterminated. The significant destruction of fish and other populations in Lake Baikal (U.S.S.R.) illustrates the effect of considerable pollution on a relatively stable body of water.

The concept of *mixing time* is employed to describe the rate of mixing of any material in a body of water or of any other ecological system. It is defined as the length of time required for a given material to become randomly distributed in the portion of the environment chosen for observation. This type of calculation (or measurement) tends to assume that the observed region can be regarded as a sort of washtub that is stirred either slowly or rapidly, as the case may be. In other words, it is a simplification based on the notion that the ecosystem approaches homogeneity. Sometimes one defines the region so that this assumption is nearly true. Sometimes one simply refuses to worry about it—a refusal often rendered necessary by the need to draw any conclusion at all.

The mixing time for a river is the time required for randomization of the material through any "slice" of the river—a cross section from bank to bank and top to bottom. This mixing time might be of the order of a few hours. The river finally bears its burden of pollution, or some part of its burden, to the ocean where a new mixing process must be considered. In this case and depending on how it is defined, the mixing time will be a matter of years, perhaps thousands of years. It is appalling to consider that molecules dumped in the ocean as a result of industrial pollution have scarcely begun to mix evenly in the sea. Even if the human race becomes virtuous (or simply disappears) and stops dumping these particular molecules, the mixing process will continue. Centuries hence whole populations of the deep ocean may be poisoned out of existence, victims of an extinct technology.

In order to go beyond the "washtub" oversimplification, the structure and dynamics of an ecosystem must be studied in great detail. In the case of the ocean, pathways must be defined by which polluting materials are transferred from one part to another. For example, if we follow a hypothetical insecticide from the field where it was applied to crops to the river and then to the sea, we discover a route that could hardly be called random mixing. The river, being less dense than seawater, contributes its water and its hypothetical insecticide to the surface layers where the insecticide might be absorbed by a primary producer such as a planktonic alga. This might, in turn, be devoured by one of the primary consumers such as zooplankton and then recycled up the food chain. In the instance where one of the planktonic organisms dies and is not eaten right away, its corpse will fall through the water into lower layers where it will be consumed by organisms of the abyss, or it will reach the bottom, contributing to the sediment. If the organism had carried our insecticide in its downward journey, it would have effected its transfer either to the populations living in the abyss or to the sediment itself. To the extent that the insecticide reaches the sediment for burial, its mixing time enters the geological time scale. Thus, barring cataclysmic events on the sea bed, this insecticide is no longer a threat to the living populations of the sea, having become effectively fossilized. A similar transfer of the chlorinated hydrocarbon insecticides to the sea bed has been shown to occur by means of absorption of the compounds to the surface of bacteria and even small inorganic particles, which then fall slowly through the water to the lower depths.

## Ecological Effects of Pollution

Pollution influences ecosystems primarily in two opposing manners: it can reduce populations or it can promote their increase. In the first instance, the substances comprising pollution can kill or sicken organisms so that entire populations are often eliminated, gaps are produced in the structure of ecosystems, and secondary disturbances occur as other organisms adjust to the new situation. Thusly one can differentiate between primary and secondary harmful effects. When, for example, DDT is sprayed on a crop, the primary effect will probably be the almost total destruction of major pest species of insects, while the secondary effects might include changes in the nematode population of the soil and the population explosion of a whole set of new, DDT-insensitive insects moving in to fill the ecological slot left by the populations exterminated. There are several examples of hitherto harmless species of insects becoming major pests as their competition is eliminated by the use of insecticides. One of the severe limitations on pesticide use is our present inability to make accurate predictions about the chain of events that will follow eradication of the target population.

A special case of secondary ecological effects is exemplified in the selective power of insecticides and other pesticides. If a harmful chemical is introduced into the environment, its ability to kill places a dramatic selective pressure favoring any members of the target population that are, for some reason, immune to its action. Thus, if a mosquito mutates to a form that is resistant to a particular concentration of DDT, then that mutant will possess an immense advantage. Its progeny will take control of the ecological niche as the unmutated, or "wild type," organisms disappear. Thus, pesticides and other introduced materials almost inevitably lead to the selection of target organism mutants that are increasingly resistant, and the resistance usually increases in a stepwise fashion as increasing doses of pesticide are required. For instance, many mosquito control projects have become increasingly difficult owing to the development of strains of the insect resistant to relatively inexpensive insecticides such as DDT. This resistance phenomenon is strictly analogous to the serious matter of pathogenic bacteria becoming resistant to antibiotics which, in some cases, are becoming useless for certain specific diseases, such as the many staphylococcus infections that are undeterred by penicillin.

In the case of both antibiotics and pesticides, the proper strategy for avoiding resistance is twofold. First, one can use such an excess of the agency that literally no organisms in the population survive; hence, there is no focus for a resistant population. In the early days of antibiotics, physicians often used just enough of a substance to produce the desired result since information about side-effects was limited. This was, of course, a good way to insure that a resistant population would be selected for, and most physicians now employ large excesses of antibiotic or a combination of antibiotics to avoid the problem. This sort of option is somewhat limited in the case of insecticides since large excesses of the insecticides produce many harmful effects on other than target organisms.

An additional way to avoid the resistance problem is to use a pesticide which is so harmful to the target organism that there are no possibilities for mutation to resistance. Thus, a mosquito is unlikely to give rise to a mutant resistant to being squashed with a hammer, although one notes that a certain selective pressure might develop favoring faster flight. A more practical example is seen in the case of the fumigation of greenhouses. Insect pests in greenhouses frequently become highly resistant to insecticides partly due to the favorable condition for their rapid growth. Therefore, the eventual recourse is to fumigate the greenhouse with a gas so toxic that it kills not only insects but all animals, including the owner of the greenhouse if he were so careless to breathe it.

A second manner in which the materials of pollution influence ecosystems is by promoting growth of some organisms rather than killing them. The manner in which phosphate harms lakes by stimulating uncontrolled growth of algae is well known. In fact, many other forms of pollution have essentially the same effect producing, this time, selective pressures *favoring* organisms that can use them as nutrients. The great adaptability of bacteria with regard to energy and carbon sources enables them to be the organisms most often favored, and the initial effect of many forms of pollution is production of a "bloom" of those types of bacteria able to prosper on the polluting materials provided. Likewise, when there is an extermination of some species owing to excessive levels of some polluting substance, it is primarily the bacteria that benefit since many are able to grow on components of the dead organisms, thereby contributing to their dissolution.

Finally, when microorganisms grow rapidly in response to pollution they obtain energy by utilizing either the offending material itself or the components of organisms already dead. Such oxidations obviously consume oxygen, and a bacterial bloom may have the effect of depleting oxygen to the extent that other organisms are unable to live even if they have survived the pollution itself. Thus, waters that are heavily polluted with organic wastes are often very nearly anaerobic (non-air-requiring), and the final act in such a series of events can be the replacement of most of the aerobic (air-requiring) bacteria with those that can live without oxygen.

## Biological Cycles of Matter

Barring the influence of man, the world ecosystem is characterized as being relatively constant in composition. Thus, the amount of carbon dioxide in the atmosphere or the planktonic structure of the sea remains roughly the same so that one may assume that there are mechanisms in the world which tend to oppose abrupt changes. Excellent examples are the *nutrient cycles* which maintain a constant chemical composition in the various parts of the ecosystem. For instance, the carbon cycle in its simplest outline forms a closed loop (see Figure 2-2) wherein animal respiration leads to the formation of $CO_2$ by oxidation of organic compounds by molecular oxygen. This half-cycle can be summarized:

$$\text{organic compounds} + \text{oxygen} \longrightarrow \text{carbon dioxide} + \text{water}$$

where the organic compounds include all of the animal nutrients such as sugars, amino acids, etc. The other half of the cycle is mainly presided over by green plants and involves the "fixation" of carbon dioxide, which means its conversion into organic compounds like sugars, amino acids, etc. This half-cycle is largely the result of photosynthesis and may be summarized:

$$\text{carbon dioxide} + H_2O \xrightarrow{\text{[light]}} \text{organic compounds} + \text{oxygen}.$$

Thus, the cycle is closed and largely self-regulating since aside from the result of the burning of fossil fuel, the concentration of carbon dioxide in

16    Chapter 2

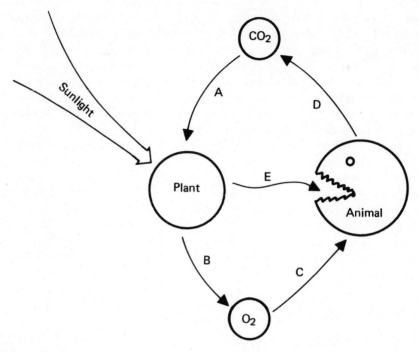

**Figure 2-2.** Simplified view of the carbon cycle. The plant conducts photosynthesis, a process leading to the incorporation of carbon dioxide ($CO_2$) from the environment into cellular organic carbon compounds. These include carbohydrates, proteins, fats, etc. This $CO_2$ incorporation is denoted by arrow A. Photosynthesis in higher plants also leads to oxygen production (arrow B). The oxygen is then used by animals to oxidize foodstuffs by the process known as *respiration* (arrow C). Respiration, in turn, yields $CO_2$ as a product, which enters the environment and may be used by plants. Finally, the animal eats the plant (or eats an organism that has eaten a plant) thereby incorporating the organic carbon which the plant has produced photosynthetically. This last process is denoted by arrow E (as well as by the gaping jaws), and it provides the foodstuffs used by the animal in respiration.

the atmosphere has remained constant. In fact, the cycle does have the capacity to absorb a good part of the excess carbon dioxide produced by industry, although we have recently gotten rather ahead of it.

Similar closed and self-regulating cycles operate also for sulfur compounds, phosphorus compounds, and nitrogen compounds, although these tend to be considerably more complex than the carbon cycle in Figure 2-2. We have just seen that industrial air pollution—production of excess carbon dioxide—can overtake the carbon cycle, and we shall see in a later chapter that other forms of air pollution can interfere with the nitrogen cycle. However, such interference is not our reason for mentioning the cycles here. Rather, these examples of closed cycles are interesting in that they provide a contrast to what occurs when novel compounds are deposited into the environment as the result of human activities. Thus, when a compound like DDT enters the natural world, there is no other half of a cycle ready to return it to its place of origin. Most matter in the environment circulates as a requirement for continuance of life on earth, but man's cleverness has led to materials which don't and can't. Part of the usefulness of compounds like DDT resides in the very low rates at which they can be degraded by bacteria and other organisms. They are not, to use the popular term, *biodegradable*, which means that their entry into the environment is essentially noncyclic. Bacteria simply cannot mutate fast enough to compete with the successive victories of the synthetic chemists.

Finally, the scale of flow of matter through one of the natural biological cycles is generally of much greater magnitude than the injection of pollutants into the ecosystem. For instance, the fixation of nitrogen in a fertile region of the earth might amount to a number of tons of nitrogen per square mile per year. On the other hand, a serious contamination of the same area with some insecticides might constitute dispersal of a few pounds or even less of the compound. Thus, pollution is frequently minor from the viewpoint of mass added to the ecosystem, but it becomes major owing to the concentration of harmful material in the cells of affected organisms and to the frequently high toxicity of small amounts.

## The Complexity of Pollution Ecology

Ecology is a field which although remarkable is also complicated. It is unfortunate that many people are led to the study of ecology by a desire to live and think simply, only to discover the fearsome complexity of its problems. Rather than a garden of love and delight, the world of ecology tends to take a mathematical turn, and the frequently expressed hope that the present ecological crisis can be solved if only we live in a virtuous manner is probably unfounded. Consider the matter of pollution. There would appear to be no single answer to the question of how to minimize its impact. We simply cannot stop polluting and much of what is called pollution is really the result of the relatively recent dominance of one species—namely us—on this otherwise biologically well-regulated earth. We are unlikely to try to reverse this dominance, although we may manage it unintentionally.

There is thus no substitute for being well informed about ecology so that we can make intelligent judgments about the probable effects of our activities. Moreover, each activity—each form of pollution—will have to be considered as a separate case but with expanding effects.

Finally, even though we are knowledgeable of a particular form of pollution and its influence on the world, we may yet be taxed as to the wisdom required to control it. Take, for example, the matter of energy. All conscientious people know that recycling is preferable to indiscriminate dumping and that aluminum beer cans should be gathered and melted down. Application of this idea should lead to a diminution of the progress of floating beer cans down our rivers to the sea, but energy is, of course, required to do the melting (*and* the collecting *and* the educating). Where will the energy be found? Obviously at the expense of another form of pollution, whether air pollution over a great coal-fired power station or, perhaps, pollution by radioactive wastes from a power reactor. How shall we assign values to the different forms of pollution so that we can make the most reasonable judgment as to the minimization of harm? It is to be feared that this central question will become more difficult to answer as we learn more about the complexity of pollution ecology.

# Chapter 3

# Cells and Basic Physiological Systems

Since this book is about the influence of pollution on cells and cellular processes, we will begin with an introduction to cell biology. The reader who has already studied this area will recognize that this is a rudimentary introduction, but it will nonetheless provide a basis for considering the central topic of this book. A reader desiring more complete information about cells is directed to the selected references in Chapter 9. It is hoped that readers who were led to the book through their interest in environmental matters will discover the delights of cell biology and will recognize that the study of life is not compartmented and must be viewed from as wide a standpoint as possible—from the subcellular to the global level.

## The General Structure of Cells

Our discussion of cellular structure will be largely concerned with animal cells since the effect of pollution on animals is of more primary interest than its effect on other forms of life. A schematic diagram of a generalized animal cell is shown in Figure 3-1. Such a diagrammatic cell is, of course, not an accurate picture of any structure found in nature except in the most conceptual way. Real cells, which are found in liver, blood, muscle, or some other specialized tissue, possess special modifications related to their location and function. The model cell shown in Figure 3-1, however, includes those structures that are most widely found in animal cells—a sort of "consensus" of cellular anatomy. For instance, virtually all cells possess a central *nucleus*, which is perhaps the most prominent feature of cells when viewed through the light microscope. Moreover, that portion of the cell which is not nucleus —the part outside the nucleus but inside the plasma membrane surrounding the cell—is termed *cytoplasm*. The older term "protoplasm" refers to the entire living content of the cell including both nucleus and cytoplasm.

A major feature of the animal cell and, indeed, all cells, is the *plasma membrane* that surrounds it. This membrane delineates the inside and outside of the cell and thus prevents the contents from mixing freely with the

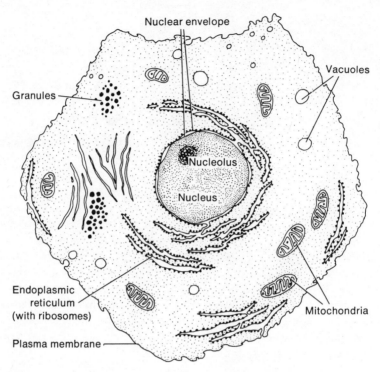

**Figure 3-1.** Diagram of an animal cell. (From John L. Howland: *Cell Physiology,* Macmillan Publishing Co., Inc., 1973.)

surrounding environment. Obviously, any harmful material entering the cell must traverse the plasma membrane, and that membrane is, in a sense, the first line of defense against such invasion. In later chapters our interest in the plasma membrane is more specific when, for example, insecticides are discussed. We will discover that compounds often have very specific and harmful effects on the plasma membrane, interfering with its ability to transport or discriminate against important compounds in the surrounding environment.

Not only is the cell surrounded by a membrane, but there are a number of important discrete membrane systems within cells. Much of the cytoplasm is packed with dense arrays of membranes in the form of closed organelles, or intracellular organs, such as mitochondria, lysosomes, chloroplasts (in green plants), and the Golgi membrane system. Much of the cytoplasm that is not occupied by these organelles is filled with an ornate system of membranes and tubules called the *endoplasmic reticulum.*

Before describing each of these structures, one final general comment should be made. Because the cytoplasm contains organelles which are closed

Cells and Basic Physiological Systems 21

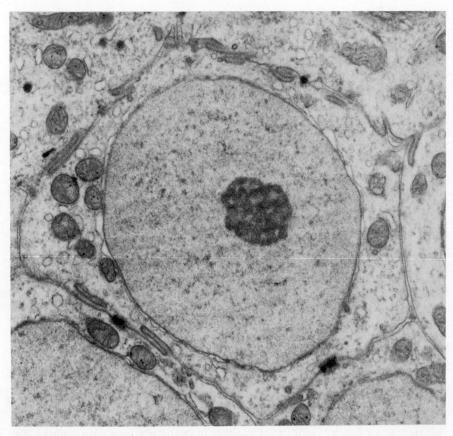

**Figure 3-2.** Electron micrograph of the nucleus of a mouse intestinal cell (magnification is × 15,000). (Micrograph by M.S.C. Birbeck, in E.J. Ambrose and Dorothy Easty, *Cell Biology,* Addison-Wesley, 1971. Used with permission from Thomas Nelson & Sons, Ltd., London.)

and therefore possess a surrounding membrane completely separating their interior from the remainder of the cell, cells must be considered as being divided into a number of compartments. Clearly, the movement of any chemical between these compartments depends upon its ability to travel across the membranes that define the compartments, and it is just at this level where much cellular regulation occurs. For instance, in the course of cellular respiration certain compounds must traverse the mitochondrial membrane from the outside cytoplasm to the mitochondrial interior where they react with respiratory enzymes. The rate of respiration and, secondarily, the rate of virtually every other cellular activity that depends on respiration are limited by the rate at which these particular compounds can penetrate the mitochondrion. Any compound that alters the ease of penetration will have a profound influence on the lives of cells. It will be evident in a later chapter that the strong affinity between many insecticides and biological membranes provides an ominous clue as to where one might look for the basis of cellular damage by such materials.

## The Nucleus

Returning to Figure 3-1, we see in the central part of our cell a large, prominent nucleus which is surrounded by an envelope consisting of two membranes, closely packed together, with round structures called nuclear pores providing communication across them. Although they resemble portholes in the envelope, these pores are actually regions in the membranes which are different in structure; they are probably modified for translocation of some substances that pass between the nucleus and the cytoplasm. Inside the nucleus one sees splotchy material termed *chromatin,* which reacts with stains that are specific for deoxyribonucleic acid or DNA. This material is actually a complex of DNA + protein[1] and, at the time of cell division (mitosis), condenses to form the chromosomes. The DNA of chromatin contains the genetic information of cells which provides the blueprint for manufacturing cells of the subsequent generation. More explicitly, the DNA in the chromatin provides the fundamental pattern for the synthesis of the proteins of the cell, a process to which we later return in detail.

Within the nucleus there is a structure called the *nucleolus* which, on the basis of specific staining as well as other chemical tests, is composed of ribonucleic acid or RNA, a compound also concerned with protein synthesis. It has recently become evident that the role of the nucleolus is related to the synthesis of ribosomes, which are particles in the cytoplasm consisting of about one-half RNA.

## Mitochondria

Mitochondria are found in the cytoplasm of most plant and animal cells and are the locations of the reactions of respiration and the aerobic formation

---

[1]. Protein is a class of large molecules composed of long chains of subunits, the amino acids. Proteins are important building blocks of cell structure as, for instance, membranes are about one-half protein and one-half fats (lipids). Moreover, all enzymes, which catalyze (cause or accelerate) biological chemical reactions, are proteins. DNA is also an example of a large molecule (molecular weight in the millions) which has as its subunits a combination of organic molecules called *nucleotide bases* plus sugar plus phosphate. DNA is located primarily in the nucleus and represents the carrier of genetic information in cells, the information being encoded in the sequence of subunits.

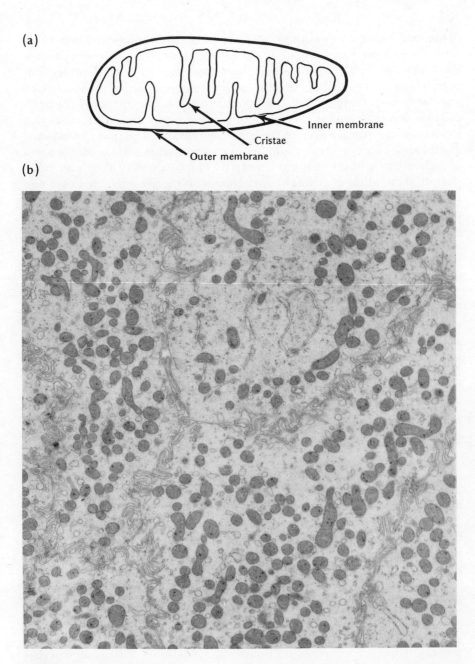

**Figure 3-3 a and b.** (a) Diagram of a mitochondrion shown in cross section. (b) Mitochondria in epithelial cells of a rabbit (magnification is × 12,000). (With permission from Dr. Gordon Kaye, College of Physicians and Surgeons, Columbia University.)

of the high-energy compound ATP (adenosine triphosphate).[2] Their structure is shown in Figure 3-3a. They are surrounded by two membranes, the inner one being deformed to form a number of inward projections called *cristae*. Since the inner membrane is partly composed of proteins which are enzymes carrying out the reactions of respiration and energy transformations, it is reasonable to assume that the cristae are modifications serving to increase the surface area of the inner membrane. The inner membrane can therefore carry a greater number of these enzymes and the efficiency of mitochondria is improved in this regard. The function of the outer mitochondrial membrane is somewhat enigmatic. It carries only a few enzymes unconnected with respiration and fails to serve as a barrier against the free diffusion of most molecules. Since mitochondria are highly specific as to what molecules and *ions* (electrically charged atoms or molecules) can penetrate their interior, the barrier that discriminates between the various possibilities must be the inner one. We shall consider transport across it in more detail later.

A final aspect of mitochondrial structure arises when we consider the inner membrane as viewed under extremely high magnification with the electron microscope. Under proper conditions, the inner membrane can be seen to bristle with numerous minuscule projections which look like lollypops. These projections have several names depending upon the viewpoint from which they are approached, perhaps the most common one being "coupling factor" or "ATP synthetase." These are the sites of ATP synthesis. Since ATP is the final product of respiration as well as the chief reason for the existence of mitochondria, the location of the projections on the inside of the inner membrane is very significant.

## Chloroplasts

The chloroplasts are important organelles found in cells of green plants. They share a number of interesting features with mitochondria, such as that both are concerned with energy-trapping by cells. In the case of chloroplasts, the energy is received in the form of visible light and is converted into chemical bond energy through the process called *photosynthesis*. Virtually all life on earth depends either directly or indirectly on this process so that any feature of pollution interfering with it on a global scale would pro-

---

2. Respiration, which is considered in detail later, may be described for present purposes as the oxidation of molecules derived from foodstuffs. This means that electrons are allowed to pass from the molecules to oxygen and that ATP synthesis is, in a manner not yet understood, coupled to this electron movement. ATP is the major high-energy compound of organisms in that its breakdown provides energy for many cellular events including movement, synthesis of many compounds, and production of heat.

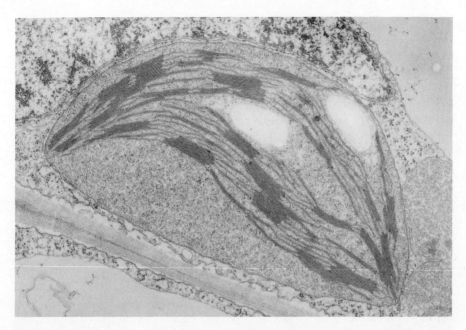

**Figure 3-4.** Example of chloroplast structure (magnification is × 17,000). Note the closely packed internal thylakoid membranes. (Micrograph by A. Greenwood, in E.J. Ambrose and Dorothy Easty, *Cell Biology*, Addison-Wesley, 1971. Used with permission from Thomas Nelson & Sons, Ltd., London.)

duce the most serious consequences, leading ultimately to a major threat to many life forms.

The membrane part of the chloroplast is the site of both the absorption of light energy and its conversion into chemical energy, just as the mitochondrial membrane is where mitochondria carry out energy-conserving reactions. The chloroplast is surrounded by a closed membrane which, as in the case of mitochondria, determines what compounds can penetrate the chloroplast interior. The inside of the chloroplast also contains a system of closely packed membranes called *thylakoid membranes* which contain the light-trapping pigments, including the chlorophyll that gives green plants their color (see Figure 3-4). Thylakoid membranes also contain the enzymes concerned with the photosynthetic production of ATP. Indeed, recent studies have shown that chloroplast membranes possess ATP synthetase projections resembling those of mitochondria. Recent biochemical studies stress the many similarities between the synthesis of ATP associated with respiration and that associated with photosynthesis; current theories devised to account for photosynthesis generally apply to most aspects of respiration as well.

Both mitochondria and chloroplasts have a system for the synthesis of proteins which is, significantly, outside the control of the nuclear DNA.

Therefore, both contain their own DNA which, like that of bacteria, is in the form of a molecular loop;[3] they also contain other components related to protein manufacture. Thus, at least some of the proteins synthesized in these structures are organized around information contained in their own DNA and not that of the nucleus. This sort of autonomy, combined with similarities between protein synthesis in chloroplasts and mitochondria on one hand and bacteria on the other, has led to the suggestion that both mitochondria and chloroplasts originally evolved as independent microorganisms which invaded the larger precursors of animal and plant cells and entered into a symbiotic (advantageous to both) arrangement with them. It is theorized that these microorganisms assumed control of the energy-conserving function of the higher cells in return for other metabolic services, such as synthesis of many proteins, which the latter provided.

### Endoplasmic Reticulum and Ribosomes

Returning to animal cells, we noted earlier that their interior contains a dense array of membranes called the endoplasmic reticulum. This intricate set of structures appears to be continuous with the nuclear envelope and exists in the cell in two forms: rough, that is, covered with small particles called *ribosomes,* and smooth where ribosomes are lacking. Much of the endoplasmic reticulum (hereafter abbreviated ER) appears to be tubular and to have the function of storage and transport within the cytoplasm. The identity of the material that is stored and transported may be deduced from the close connection between rough ER and the bound ribosomes which are the sites of cellular protein synthesis. For example, cells that are especially active in making and secreting proteins (as in the cases of cells that make either digestive enzymes or protein hormones such as insulin) have very rich concentrations of both ribosomes and endoplasmic reticulum. In such cells an additional set of structures is seen to be highly developed, namely the *Golgi complex,* which is a set of flattened sacs near the nucleus and which appears to be a collecting point for proteins prior to secretion. Although all of these structures are wonderfully clear and complex when viewed with the electron microscope, in the living cell, which cannot be examined with an electron microscope,[4] the structures are amazingly fluid and transitory, forming and breaking down with great ease and rapidity. One should always be aware of the dynamic character of the interior of cells, especially when examining diagrams (such as Figure 3-1) which on the printed page are as still as death.

---

3. DNA should be thought of as a long chain composed of two strands twisted together (a helix). In bacteria, mitochondria, and chloroplasts, the chain is in the form of a loop, i.e., it has no free end.

4. Electron microscopes are only generally able to view cells that have been specially "fixed" (i.e., killed) and stained.

Cells and Basic Physiological Systems   27

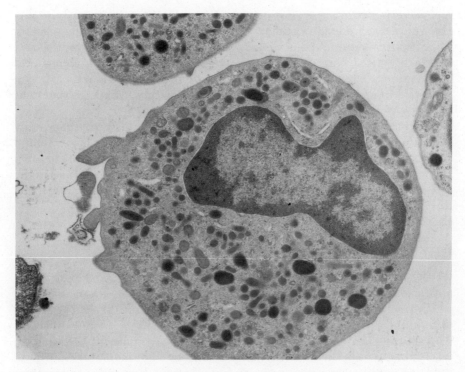

**Figure 3-5.**   White blood cell from a guinea pig. Note the many lysosomes in the cytoplasm. (9500 ×, from Fawcett, D. W.: *The Cell: Its Organelles and Inclusions.* Philadelphia, W. B. Saunders Company, 1966, p. 201.)

The ribosomes perching on the ER (making it rough ER) have, themselves, a structure that has been extensively studied and is closely related to their function as sites of protein synthesis. They are composed of two parts, one somewhat larger than the other, which may be separated by placing a suspension of ribosomes in a medium with a low concentration of certain salts. As previously stated, the ribosomes are the sites of cellular protein synthesis and are about one-half protein with the remainder being ribonucleic acid (RNA), a long chain molecule quite similar to DNA.

## Other Cellular Structures

There are a number of other membrane-formed organelles. Many cells contain *lysosomes* which are somewhat spherically-shaped organelles surrounded by a membrane (see Figure 3-5). They contain a variety of enzymes that are involved in intracellular digestion of foods since they break down such materials as large sugars, fats, proteins, and DNA. It is obviously important that the cell not allow contact between these enzymes and the protein, DNA, RNA, and other molecules which form the cell itself; the result would be self-digestion. Thus, the enzymes are sequestered in the lysosomes, and the function of a lysosomal membrane is to regulate their action. Clearly, the lysosomal membrane is immune to their activity or it would be digested; the basis of such immunity is not clear.

Other configurations of membranes within the cytoplasm are sufficiently common and have an adequately characteristic structure to be given names. An example is the Golgi complex which, as we noted earlier, appears to function as a collecting point for proteins about to be secreted by cells, as in the case of cells that manufacture digestive enzymes. It may also play a role in the formation of lysosomes.

## Basic Topics in Cellular Chemistry

Of the many classes of chemical compounds occurring in cells, two will appear prominently in the remainder of this book. The first is the proteins, which are of great importance in the structure of cells and, as we mentioned, include the enzymes which catalyze all of the chemical reactions of cells. The second class of molecules includes the fats or lipids, which are most interesting to us in their role as a major component of membranes. We will see later that certain components of pollution are harmful through their destructive interaction with enzymes while others, notably the insecticides, act on membranes and interact in a specific way with the membrane lipids.

### Proteins

Proteins are composed of smaller subunits called the *amino acids* which are named for their two chemical groups, the amino group, $-NH_2$, and the acidic carboxyl group, $-COOH$, which give them many of their characteristic properties. An example of an amino acid is alanine which has the following structure:

$$CH_3 - \underset{\underset{H}{|}}{\overset{\overset{NH_2}{|}}{C}} - COOH$$

Approximately twenty amino acids are commonly found in proteins, and they mainly differ only in the nature of the group that is a $CH_3$ in our example, alanine. A given protein might contain a selection of a dozen or more different amino acids, and it might contain a total of one hundred individual ones. The properties of a particular protein—for instance, the nature of its action as an enzyme—depend entirely on the composition and order of the individual amino acids that comprise it. Much of protein chemistry in recent years has been a matter of determining the amino acid sequence of proteins and attempting to relate that sequence with matters of function.

Obviously, proteins with a hundred subunit amino acids are very complex in structure, but one simple statement can be made about them: all amino acids are connected by the identical linkage regardless of what they are or where in the protein structure they occur. The fundamental bond connecting amino acids in proteins is called the *peptide bond,* and it connects them through the amino group of one and the acid (carboxyl) group of the next. To understand this bond, imagine that the two groups are combined and that water ($H_2O$) is somehow removed.

$$-NH_2 \quad HOOC-$$

After the water is "extracted" the remaining atoms are left linked in the following manner:

$$-\underset{|}{\overset{H}{N}}-\underset{}{\overset{O}{\overset{\parallel}{C}}}-$$

This configuration is the peptide bond. The chemical mechanism for the formation of the bond in cells requires a complex set of enzymes (which, of course, contain many peptide bonds themselves), and it occurs at the ribosomes. Water is really removed in the reaction—i.e., it is one of the products of the reaction. When peptide bonds are broken (a task also performed by specific enzymes in cells), the reaction is *hydrolysis* which is the addition of water leading to bond cleavage.

The exact properties of proteins depend partly upon the manner in which the long chain of amino acids is convoluted in space. The different amino acids have different abilities for forming weak bonds with other amino acids; these bonds determine the conformation of the protein, that is, its three-dimensional structure. For instance, some proteins are largely coiled like a spring; others, including many enzymes, are globular with the amino acid chain wrapped about itself; and still others, such as B keratin, form flattened sheets. Changes in the shape of these huge molecules are often associated with changes in function such as enzymatic activity. An important part of cellular regulation is involved with the transformation of enzymes from an active to an inactive configuration and vice versa.

Finally, we said that different amino acids have different groups located where the $CH_3$ group was in alanine. This means that at different points, proteins expose these different groups to the outside medium and many of

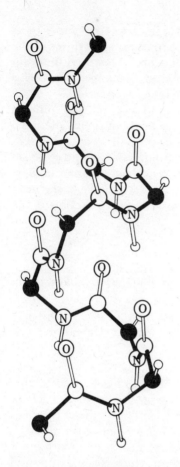

**Figure 3-6.** The fundamental helical (coiled) structure underlying many proteins. (From William B. Wood et al, *Biochemistry: A Problems Approach,* copyright © 1974, W. A. Benjamin, Inc., Menlo Park, California.)

them have special implications for the function of the protein. Thus, the amino acid glycine has only a hydrogen atom in place of the $CH_3$ while other amino acids have a variety of functional groups including the following:

$$\begin{array}{c} CH \\ \diagdown \\ CH_3 \diagup CH- \end{array}$$ (valine)

and $-CH_2OH$ (serine)

and ⟨phenyl⟩$-CH_2$ (phenylalanine)

and $-CH_2SH$ (cysteine)

Both cysteine and serine play a central role in the catalytic function of many enzymes, and both are targets of certain materials that are harmful to cells. For example, one of the effects of heavy metals, such as mercury, on cells reflects its binding to the $-CH_2SH$ groups of proteins and the consequent inhibition of the enzymes' activity.

## Lipids and Membranes

Lipids are a varied group of compounds which have in common their low solubility in water and high solubility in a variety of materials that are often called lipid solvents, which include various petroleum fractions, ethers, and other organic compounds. This definition of lipids sounds circular, and it is. In the context of this book we don't have to be very concerned about an inclusive definition for the term since the few lipids of interest to us in our present study are easily defined. We are interested in lipids solely in the context of their role in membranes. For our purposes, membranes contain two primary types of lipid, as well as many minor constituents and large amounts of protein. If membranes are roughly one-half lipid, then about one-half of this substance is composed of a class of lipid called *phospholipid*. Membranes from higher plants and animals (but not bacteria) also contain cholesterol which has the following structure:

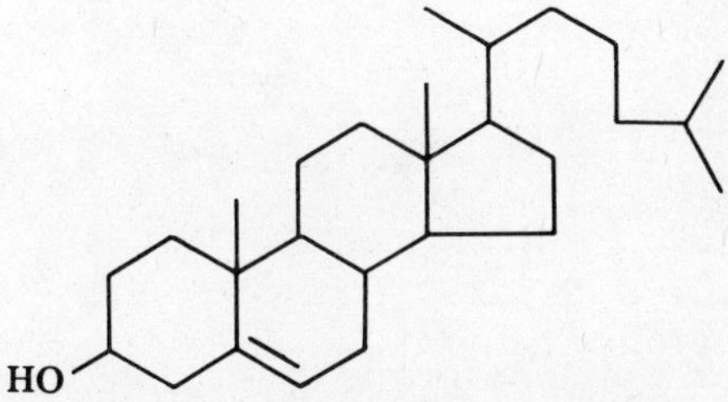

[Structure of cholesterol with HO- group]

Cholesterol has been much discussed in the popular press in regard to its apparent role in circulatory diseases.

Phospholipids are, as their name indicates, lipids which contain phosphate. Their general structure is fundamental for understanding lipids and is built on the following pattern:

1. Many lipids are composed of the alcohol glycerol

$$\begin{array}{c} CH_2OH \\ | \\ CHOH \\ | \\ CH_2OH \end{array}$$

which is linked to as many as three long-chain acids called *fatty acids* of which an example is $CH_3CH_2CH_2CH_2CH_2CH_2CH_2CH_2CH_2CH_2CH_2$ $CH_2CH_2CH_2CH_2CH_2CH_2COOH$. Note that one end of this very long molecule bears the acidic carboxyl group —COOH, just as in the case of the amino acids. In fact, the combination of these compounds with the alcohol glycerol occurs with the removal of water in a manner analogous to the formation of the peptide bond of proteins. Thus, one fatty acid attached to glycerol would have the following appearance:

$$\begin{array}{l} \quad\quad\quad\quad O \\ \quad\quad\quad\quad \| \\ CH_2O-C-CH_2CH_2CH_2CH_2\ldots \\ | \\ CHOH \\ | \\ CH_2OH \end{array}$$

Such compounds are called *monoglycerides;* when there are three fatty acids—one attached to each —OH group—the compound is called a *triglyceride*. Triglycerides, which are common components of cells, are what people usually mean when they refer to "fats."

2. Phospholipids are very similar to triglycerides except that one of the fatty acids is replaced by a phosphate.

$$\begin{array}{l} CH_2O-(\text{phosphate}) \\ | \\ CHO-(\text{fatty acid \#1}) \\ | \\ CH_2O-(\text{fatty acid \#2}) \end{array}$$

Indeed, other groups can also be attached to the phosphate. An example of a common class of phospholipids is the phosphatidyl serines where the amino acid serine is attached to the phosphate group.

$$\begin{array}{l} \quad\quad\quad\quad\quad\quad\quad\quad\quad O \\ \quad\quad\quad\quad\quad\quad\quad\quad\quad \| \\ \quad\quad\quad\quad\quad\quad CH_2-O-C-R \\ \quad\quad\quad\quad\quad\quad | \\ \quad O \\ \quad \| \\ R-C-O-CH \\ \quad\quad\quad\quad\quad\quad | \quad\quad\quad O \\ \quad\quad\quad\quad\quad\quad\quad\quad\quad\quad \| \\ \quad\quad\quad\quad\quad\quad CH_2-O-P-O-CH_2CH-COOH \\ \quad\quad\quad\quad\quad\quad\quad\quad\quad | \quad\quad\quad\quad | \\ \quad\quad\quad\quad\quad\quad\quad\quad\quad O \quad\quad\quad\quad NH_3^+ \end{array}$$

Cells and Basic Physiological Systems    33

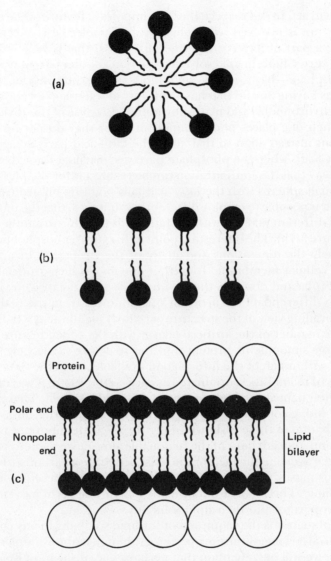

**Figure 3-7.** Interaction between the fatty acids of phospholipids leading to (a) the spherical configuration, (b) a lipid bilayer, and (c) a biological membrane which also includes proteins.

We have extensively discussed the chemistry of lipids because the structure of phospholipids is of central importance in understanding how membranes are formed and hence how compounds harmful to membranes can be thought to act. In this respect, the most important feature of the phospholipid structure is involved with solubility. The molecule is exceptional in possessing a part of its structure (the fatty acids) that is very insoluble in water and very soluble in lipid solvents, whereas another part of its structure —the phosphate—has the opposite solubility. One can consider the whole molecule as having an end that has an affinity to water and an end that does not (i.e., a hydrophobic and hydrophylic end). Because of this divided preference, when one places phospholipids in water they do not form a true solution but instead align so that all of the fatty acid parts are pointed to other fatty acids while the phosphate parts are oriented toward the water. The only way that the configuration can be realized is for the phospholipid to form small spheres with the fatty acid tails pointing inward (see Figure 3-7). Now it is possible to employ different conditions so that the mixture will orient in a different manner with the fatty acids pointing to another layer of lipids (Figure 3-7b). The interesting point is that such a phospholipid bilayer has precisely the dimensions and many of the properties of a naturally occurring cellular membrane. Indeed, if one makes such artificial bilayers in the test tube and observes them with an electron microscope, it is quite difficult to differentiate the artificial "membranes" from the real ones.

The prevailing view of the structure of cellular membranes is that they are built in the manner of the artificial bilayer with the added feature of insertion of large amounts of protein into the structure. This view, which is consistent with much of the information available about the structure and properties of membranes, originated as early as the 1930s; it was developed by (and often named for) the pioneer in this field, Danielli. There are various views today concerning how the proteins are inserted into the lipid framework; one of these is shown in Figure 3-7c. An important feature of this view (and, indeed, most of the reasonable alternative models) is that any compound traversing the membrane must penetrate a part of it (the central core) that is mostly or entirely lipid, and its passage is therefore aided if it is lipid-soluble. This is borne out by many measurements of movement with an ease proportional to their lipid solubility.

A related matter is the requirement for lipid solubility in any compound expected to affect membrane properties. Although many examples could be cited, here we will only mention that perhaps one of the most lipid-soluble compounds known is the insecticide DDT, which acts on the cellular membranes of both insects and nontarget organisms. We will return to this point at length in Chapter 5.

## Cellular Processes

The remainder of this chapter is devoted to outlines of some cellular processes which we describe prior to our consideration of effects of pollution upon them. Although these processes comprise part of cell physiology, our outline must omit many topics in this area and concentrate on a few that are most relevant to the topic of this book. A reader desiring more information about any of the topics or about the wider scope of cellular physiology is directed to the list of references in Chapter 9.

## Enzymes: Biological Catalysts

We shall encounter examples of noxious materials interfering with the normal lives of cells by inhibiting various enzymes. It is readily comprehensible how enzyme inhibition should lead to cellular disarray since virtually all chemical reactions of cells are catalyzed by enzymes.

A *catalyst* enables a reaction to proceed more rapidly than it would in the absence of this substance. A reaction that is fundamentally impossible, such as making gold from lead, cannot be made to proceed regardless of what catalyst is employed (and alchemists checked an impressive number of possibilities). In other words, a catalyst increases the velocity of a reaction without altering any other intrinsic properties of the reaction, such as the nature of reactants and products, the position of the equilibrium toward which the reaction tends, or the amount of heat that is given off or absorbed.

Since magic is no longer a reasonable explanation for physical events, we conclude that a catalyst must participate in the reaction. On the other hand, an important feature of the definition of a catalyst is that it is not used up in the course of the reaction. Thus, it must perform a cyclic role as a reactant in part of the reaction but being returned in its original state at the conclusion of the reaction. Let's consider a general case involving an enzyme (it could also be an inorganic catalyst, though) in which a cellular reaction is the following:

$$A \rightleftarrows B.$$

Like most cellular reactions, this reaction is reversible, which is the reason that the arrows extend in both directions. Furthermore, the reaction is so slow at body temperature—e.g., the cell is in a warm-blooded animal at about 37°—as to be practically immeasurable and quite useless to the cell. In the test tube one can speed up this reaction (and other chemical reactions) by increasing the temperature, but that option is of no use to the cell which cannot survive at much higher temperatures. Therefore, the cell must have recourse to a catalyst that is specific for this particular reaction. We shall call this enzyme "E" and remember that it is specific for this reaction only. As we said, enzymes must react with their reactant but must also be regenerated at the end of the reaction. Therefore, the more detailed reaction should be written:

$$A + E \rightleftarrows A{-}E \rightleftarrows B{-}E \rightleftarrows B + E.$$

Thus, the actual conversion of A to B occurs with the molecule bound to the enzyme, E. At the end, B leaves the enzyme so that there is a free molecule of E to carry out the reaction with another molecule of A. This is what we mean when we say that the role of the enzyme is cyclic: one molecule of enzyme can react repeatedly with molecules of A. In fact, there are enzymes with as many as one million molecules of reactant per second, which means that the cell does not require many molecules of a given enzyme to catalyze a reaction. This is just as well, for enzymes are proteins and are therefore huge molecules which would tend to fill up cells if too many copies were required.

Much effort has been expended in trying to discover how the binding of "A" to an enzyme makes it more likely to react to form "B." One theory is that there is a sort of energy barrier between reactants and products in any chemical reaction. Energy required to surmount the barrier can be provided in the form of an increased heat which is why reactions are speeded up by an increase in temperature. On the other hand, the presence of a catalyst such as an enzyme lowers the barrier so that less energy is required to overcome it. Thus, the reaction proceeds at a greater velocity at a lower temperature. This explanation does provide some insight into enzyme function but also, in a sense, it begs the question as it does not explain how the binding of a reactant to an enzyme lowers the energy barrier. The answer to that question remains rather incomplete, but we can summarize much of what is known about the process as follows:

1. The reaction between an enzyme and a reactant (sometimes called a *substrate*) occurs at an *active site* on the enzyme, and the active site comprises a small fraction of the total surface of the whole enzyme. It is the shape of this site that gives the enzyme its specificity—that enables it to react with A but not Z. One can visualize a sort of lock-and-key arrangement where there is a tight fit between the active site and the small reactant.
2. The tight binding of substrate to enzyme must alter the reactivity of the substrate. This can occur through a number of mechanisms, but in all cases the chemical properties of the molecule must be altered sufficiently to increase the reaction rate manifold. For example, a catalyst can make a molecule more reactive by securing it in a precise way on the active site so that a particular reaction is favored. If A is going to combine with X and the enzyme binds A and X in close proximity, the reaction is likely to be favored over the situation where both molecules must "find" each other in free solution, relying on random movement to bring them together. In other cases, the tight binding of the smaller substrate to the enzyme can produce strains in the structure of the substrate; in general, a "strained" molecule will be more reactive in a

variety of chemical reactions. In some cases the enzyme, through its binding to the substrate, produces specific changes in the substrate such as withdrawing electrons from specific bonds or injecting electrons into specific regions of the molecule. It is likely that any given enzyme combines a number of such mechanisms in its catalytic effects and it is, in some instances, rather difficult to distinguish the contributions of different ones.

*Enzyme Inhibition.* Enzyme inhibition is of interest to us for two reasons. First, compounds that inhibit specific enzymes are harmful to cellular processes. Secondly, studies employing inhibitors have provided a great body of information about how enzymes function. For instance, a number of organic mercury compounds inhibit many enzymes apparently by binding to sulfhydryl groups (—SH) in the enzymes. One immediately learns about the harmful effects of mercury compounds and about the role of sulfhydryl groups in enzyme catalysis as well. In this case, compounds that demonstrate an especially high sensitivity to the mercury compounds might be inferred to have a sulfhydryl group playing a central role in the catalytic process, as it is possible that they are located at the active site of the enzyme and are involved with binding of the substrate.

A class of enzyme inhibitors called competitive inhibitors have attributes that are most instructive in the general study of toxic compounds. Competitive inhibitors act by "looking" enough like a substrate on the active site of the enzyme. Thus, they compete with the substrate for a place on the enzyme surface. One can shift the balance between substrate and inhibitor by adding more of one or the other of the compounds. A classical case of competition of this kind is the inhibition of the enzyme succinate dehydrogenase[5] by malonate. The normal reaction involves the removal of electrons and hydrogen from succinate

$$\begin{array}{c} COO^- \\ | \\ CH_2 \\ | \\ CH_2 \\ | \\ COO \end{array}$$

giving as the final product, fumarate.

---

5. Enzyme names denote both the substrate of the reaction (in this case, succinate) and the character of the reaction (dehydrogenase, the removal of hydrogen).

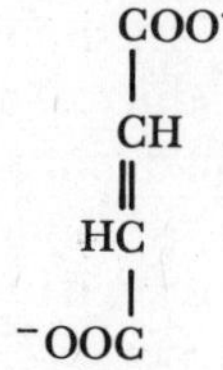

$$\begin{array}{c} COO^- \\ | \\ CH \\ \| \\ HC \\ | \\ {}^-OOC \end{array}$$

Malonate is able to compete for the active site of the enzyme due to its similarity to succinate, but the similarity is not sufficiently strong to allow the reaction to occur. Malonate

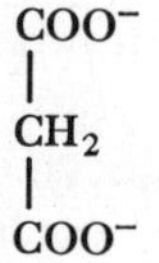

$$\begin{array}{c} COO^- \\ | \\ CH_2 \\ | \\ COO^- \end{array}$$

has the two carboxyl groups of the normal substrate but lacks an adjacent $CH_2$ group which is, in fact, essential for the reaction (which is removal of hydrogen and an electron from each of two $CH_2$ groups).

The phenomenon of similar compounds competing for active sites is widespread in biology, and it involves not only active sites of enzymes but also binding sites on membranes. Such competitive inhibition is clearly harmful in many instances, but it can also be useful from the human point of view. Many drugs employ such competition, of which a good example is a number of antitumor drugs that are close (but not too close) mimics of important cellular compounds. For example, adenine is a normal constituent of cells as it forms a starting material for the synthesis of DNA, RNA, and ATP. The structure of adenine is

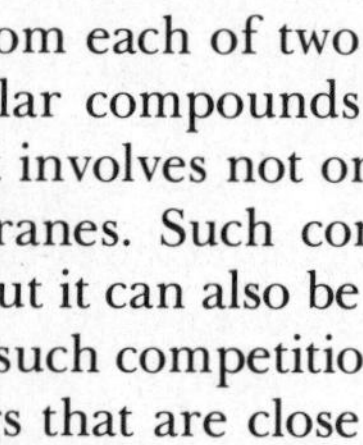

[adenine structure]

and it is not surprising that 6-mercaptopurine is an inhibitor

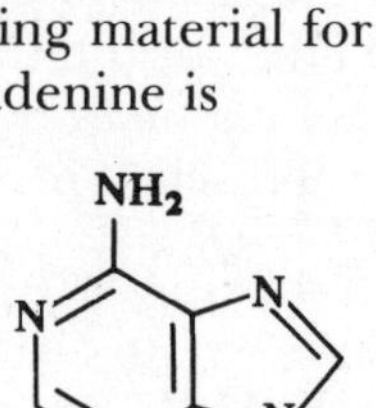

[6-mercaptopurine structure]

of DNA synthesis, among other activities. It turns out that certain kinds of tumor cells (which are making DNA at a fast pace) are more sensitive to this compound than are normal cells so that the compound exerts a selective effect, favoring normal cells over the tumor ones.

The point of this discussion is that competition between similar molecules is frequent and often harmful to cells. Therefore, if one is trying to determine the basis for harmful effects of a given compound, a very useful strategy is to consider whether the compound is structurally similar to any normal constituents of cells. Sometimes the similarities are subtle since active sites and human scientists do not necessarily "view" molecular structure in exactly the same way. It is nonetheless a very useful approach, and one can gain a certain facility in thinking like an active site. After all, there are really only two fundamental ways in which compounds (say, components of pollution) *can* affect enzymes or cells: they can either react with some part of the enzyme (or cell) and alter its properties, or they can compete with a normal reactant at an active site of some kind. It is surprisingly difficult to think of other possibilities.

## Respiration

It is a truism that cells require energy and that the only means they have devised through evolution to extract it from their environment is by dispersing electrons. More specifically, cells move electrons from a higher energy to a lower one and conserve the energy difference in the form of chemical bonds. This rather simplistic view of cellular energies nonetheless provides a very complete view; there are no energy-conserving reactions in cells that do not involve electron transfer. Since adding an electron to an atom (or molecule) is called *reduction* and removing one is called *oxidation,* we can state the same general rule in different terms: all cellular processes that function to conserve (trap) energy include oxidations and reductions. This is true for respiration, the chief form of energy trapping in animals, and for photosynthesis, the major form in plants and, of course, the primary energy trap in the whole biosphere. It is even true for fermentations, which are pathways for energy conservation in yeasts, bacteria, and other organisms where energy is obtained through chemical reactions occurring in the absence of both light and oxygen.

Before proceeding to respiration, it is of interest to see how the dogmatic assertion about moving electrons does apply to respiration, photosynthesis, and fermentations. In the case of *respiration,* the process involves the removal of an electron from a food molecule or a molecule derived from food. The electron is then passed along a chain of electron carriers until it is finally placed on a final electron acceptor which is oxygen in the case of animals and plants (but not certain bacteria). In the case of the few bacteria that are exceptions, the final electron acceptor might be the oxidized form of iron or another oxidized inorganic substance. The passage of electrons toward the final electron acceptor is coupled to the synthesis of high-energy chemical bonds (we will explain later how this occurs). The net process of respiration in animals and plants can be summarized as follows:

```
                    (electron flow)
   food molecule ----------=---------- oxygen
                              ↘
                          high-energy
                             bonds
```

When the electrons are finally passed to oxygen, the oxygen becomes reduced, picks up two hydrogen ions from the solution around it, and becomes water. This reaction can be written

$$O + 2 \text{ electrons} + 2H^+ \rightarrow H_2O.$$

Organisms that respire produce "metabolic water" which is, in part, how some animals can survive with minimal external water supply.

*Photosynthesis* operates by a quite similar process. In this case the electron source is different; the interaction of light and the photosynthetic pigment, chlorophyll, causes the pigment to emit an electron which is then able to travel along a chain of electron carriers similar to that of respiration. The emission of the electron from chlorophyll leaves an electron deficiency on the molecule. This electron "hole" may be filled by another electron from a water molecule (which is thus oxidized to form oxygen by a reverse of the respiratory reaction). Thus, chlorophyll functions as an electron pump in the presence of light.

*Fermentations* are defined as processes which do not require oxygen and have a pathway so arranged that the final electron acceptor is derived from the electron donor. For example, imagine that a yeast cell ferments a sugar by a pathway involving several hypothetical intermediates.

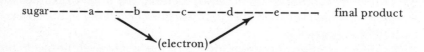

The electron is removed at an early point in the sequence and is returned later. It is likely to be carried by an electron carrier which is called a coenzyme in this instance. In what is probably the most famous fermentation reaction carried out by yeast, the final product is ethanol, $CH_3CH_2OH$, which is well known for its pharmacological and toxic properties.

These three types of energy-trapping reactions—respiration, photosynthesis, and fermentation—have one other feature in common. In all cases, the final trapping of energy is in the form of so-called high-energy chemical bonds, and specifically a particular bond in adenosine triphosphate (ATP). ATP has three phosphate groups aligned in a row, and the bond leading to the last one is the high-energy bond. This means that it requires a lot of energy to form ATP and a lot of energy is provided when it is broken. The structure of ATP may then be outlined as

adenine ring—ribose (a sugar)—phosphate-phosphate-phosphate

or written more explicitly as follows:

$$^-O-\overset{\overset{O}{\|}}{\underset{\underset{O^-}{|}}{P}}-O-\overset{\overset{O}{\|}}{\underset{\underset{O^-}{|}}{P}}-O-\overset{\overset{O}{\|}}{\underset{\underset{O^-}{|}}{P}}-O-CH_2-\text{(ribose-adenine)}$$

The reason for stressing this molecule is that the breaking of that final phosphate bond provides energy for virtually all cellular functions and reactions, including muscular contraction, amoeboid movement, light emission in certain organisms, the synthesis of many compounds, production of "body heat," and many other processes. Furthermore, as we shall see in later chapters, this molecule is in various ways a focus for many harmful effects that various components of pollution have on cells.

To return to respiration, the passage of electrons from a food molecule (such as a molecule derived from a sugar) toward oxygen occurs by way of a series of carriers called the *respiratory chain*. This chain is summarized in Figure 3-8. With the exception of the small molecule called ubiquinone, all of the carriers are proteins. Flavoproteins are yellow proteins which contain a flavine group related to the vitamin riboflavin. The cytochromes are red to greenish-red and contain a heme group with an iron atom in the center (see Appendix). Cytochromes are similar to the blood oxygen carrier hemoglobin in their possession of an iron-heme complex. In fact, it is the iron that gives and accepts electrons and gives the cytochromes their red color. Interestingly, if the iron is replaced with magnesium and some other changes are made, the heme becomes the ring system of chlorophyll which, as we saw, is another carrier serving a different but related function. Both plants and animals contain this respiratory chain; it is located in their mitochondria,

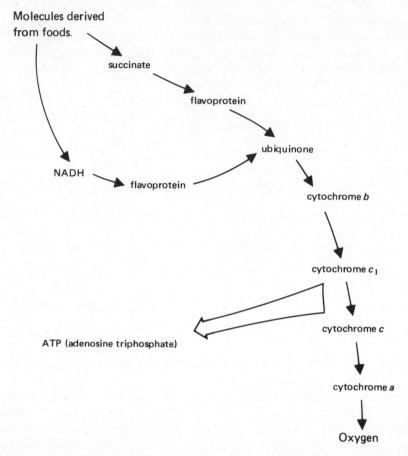

**Figure 3-8.** The passage of electrons along the mitochondrial respiratory chain. The small arrows denote electron transfer; the large arrow indicates coupled synthesis of ATP. NADH is an abbreviation for the chemical name of a small molecule that is involved in many oxidations. The chain is somewhat more complex than shown, as there are at least two cytochromes $a$ and two cytochromes $b$.

particularly in the inner mitochondrial membrane. Finally, we said that electron flow along the respiratory chain is coupled to the synthesis of the terminal phosphate bond of ATP. The astonishing fact is that, although this is one of the central reactions of the cell, we do not know how it works. The mechanism for ATP synthesis coupled to respiration is one of the great unsolved problems in cellular biology and, for the purposes of this book, we need only say that any compound which interferes with respiratory ATP synthesis will surely damage the vital processes of a cell. We shall later encounter compounds that harm cells in two main ways: either they inhibit the passage of electrons along the chain or they "uncouple" the ATP synthesis from the electron flow. A good example of the first type of compound is cyanide, which prevents the final reaction of the chain and is notoriously harmful to cells.

## The Synthesis of Proteins

We shall now discuss protein synthesis which is included for several reasons. First, this cellular process is of prime importance since proteins are major components of cells and all enzymes are proteins. It is also an example of a process that requires ATP as an energy source. Also, basic knowledge of protein synthesis will be useful in later discussions.

Protein synthesis in cells occurs in ribosomes, which are either located in the free cytoplasm or within other organelles such as mitochondria and chloroplasts. We saw earlier that proteins are composed of a number of different amino acids in a *particular* sequence. For this reason the mechanism of protein synthesis must include two main processes. First, the cell must form the bond between adjacent amino acids; this is where the energy is required. Also, the cell must specify the exact order in which amino acids are to be linked. Thus, there is an energetic problem and an informational problem.

The information that specifies the order of amino acids in a protein resides in the DNA of the cell, which is mostly in the nucleus. The DNA—the genetic material of cells—has the role, at least in the context of protein synthesis, of carrying from generation to generation the information required to specify the amino acid sequences of all the proteins in the cell. Highly regulated procedures have evolved in the cell in order to insure that exact copies of this information are transferred to subsequent generations, the best known process of which is mitosis.

Information is coded in DNA in the form of another sequence of molecules: that of the *nucleotide bases* which partly comprise the DNA molecule. Information is then transferred from DNA to a molecule called *messenger RNA* (mRNA) by means of specific base-pairing relationships. Thus, the RNA is synthesized using the DNA as a pattern so that the sequence of bases in DNA generates a specific sequence in the mRNA molecule. The mRNA serves, in turn, as the pattern for protein synthesis. The pattern is "read" in such a manner that each sequence of three bases in the RNA sequence denotes the location of a specific amino acid in a protein chain. Special sets of three bases in the mRNA tell when to start and terminate synthesis of a protein and, of course, all of this information was originally

44    Chapter 3

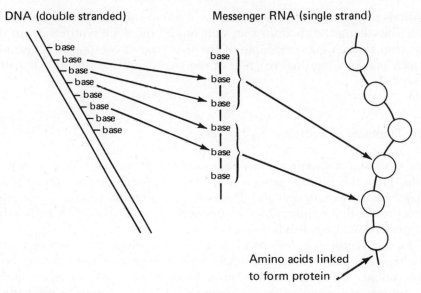

**Figure 3-9.** Information transfer from DNA to messenger RNA (mRNA) to completed protein. Note that a three-base "word" in either DNA or RNA corresponds to a single amino acid location in the protein chain.

located in the DNA chain. These relations are shown schematically in Figure 3-9.

With this understanding of the origin of the information for making a particular protein, we can proceed to the overall process of protein synthesis which occurs on the ribosome and can be divided into the following two parts:

1. Activation of the amino acids so that the peptide bonds can be synthesized and
2. The reading of the sequence information in the mRNA. The first of these processes requires ATP, the first reaction being

amino acid + ATP ⟶ activated amino acid + some leftovers followed by

activated amino acid + tRNA ⟶ amino acid—tRNA + some leftovers.

The activated amino acid is actually the amino acid attached to most of the ATP molecule (less two phosphates). This is called "activated" because of its greater reactivity than the plain amino acid. It can, for example, now react with the *tRNA molecule* (transfer RNA) which is the small RNA that will subsequently enable the amino acid to find its place on the mRNA pattern. The amino acid is still activated when attached to the tRNA as it has enough energy, so to speak, to form the peptide bond with another amino acid. Both of these reactions are carried out by enzymes located in the free cytoplasm; what follows occurs only at the ribosome.

This process involves the ribosome binding to the threadlike mRNA molecule which, as we recall, bears the sequence information. Then the ribosome slides along the mRNA and, as it slides, a protein is synthesized. This occurs because transfer RNA molecules with amino acids attached are bound to the mRNA by the same kinds of base-pair arrangements that we cited earlier so that they come to rest there in the proper order for the protein structure. One can imagine that the amino acid now bears a hook—the tRNA—which has three bases that are complementary to three at that location on the mRNA thread (see Figure 3-9). As the tRNA-amino acids bind to their proper place on the thread, enzymes on the ribosomes catalyze the final reaction leading to the formation of the bond between the adjacent amino acids. When the ribosome, sliding along the thread of mRNA, reaches one of the sequences of three bases that denote "end of protein," then the process ends. The completed protein, with its approximately one hundred amino acids all neatly linked, comes off as a finished product.

## Some Membrane Processes

We now direct our attention to some processes occurring at cellular membranes which, as we saw, not only surround cells, but also make up many of the organelles of the cell interior. In this book we shall be especially interested in membranes because many of the effects of pollution occur there and because the outside membrane of surrounding cells represents the first line of defense against foreign materials.

Membranes are interfaces between different compartments which are able to control the movement of materials between them. Many of the membranes of cells are termed "semipermeable" because they are permeable to some substances but not to others. In other words, they allow only some to penetrate so that, for instance, many cells are much more permeable to potassium than the very similar ion, sodium. In general, if other factors do not intervene, membranes will be more permeable to materials that dissolve in the lipids of the membranes, and it is significant that virtually nothing is more soluble in lipid than many insecticides, including DDT. If a membrane divides two spaces—e.g., the interior from the exterior of a cell—then material X can move from one space to the other provided that the membrane is permeable to it. If the material is highly insoluble in lipid or if it is very large, like a protein, it is unlikely that it can penetrate the membrane. If it can traverse the membrane, another set of conditions will determine whether it can actually move. For example, net movement of a molecule across a membrane (or anywhere else) requires energy of which there are two possible sources.

1. The available energy can be in the form of a concentration difference. Thus, X can move from a space where it is at a high concentration to a space where the concentration is lower. When enough of the material has moved for the concentration to become equal, then we consider the energy to be expended.
2. The energy can generate from the cell's metabolism. Usually this means that the energy can be supplied by ATP which is, as we have seen, produced in a process related to respiration (or another form of electron movement). When the energy is provided by a circumstance other than the concentration difference, i.e., from metabolic energy, we call the process *active transport*. Active transport can actually drive molecules or ions against their concentration gradient. In other words, membranes can concentrate compounds to an extent far beyond the point where there are equal concentrations on either side of the membrane. Sometimes cells extrude materials so that the interior concentration will become much lower than the outside one. For example, algae living in the ocean actively extrude sodium and absorb potassium so that their inner sodium concentration falls far below the level of seawater while the potassium concentration is considerably higher. Other marine animals concentrate some of the trace components of seawater, including rare earth ions, to the point where they are enriched as much as a millionfold over the external concentration. If the source of energy is removed by adding an inhibitor of respiration (or photosynthesis, as the case may be) then the concentrations inside will soon approach that of seawater. Active transport is thus important in regulating the internal content of marine organisms (and we shall see that insecticides are sometimes able to disturb it). Membrane transport is also responsible for many other functions in all types of cells. We shall encounter it shortly in a context that is very important to the subject of this book: the mechanism of nerve transmission.

A good example of active transport is the *sodium-potassium pump* or ATPase[6] which is responsible for nerve transmission, some features of muscular contraction, and many other events that occur at the surface of cells. This system, which has been most extensively studied in nerve and red blood cells, resides in the plasma membrane and couples the uptake of potassium and the extrusion of sodium to ATP breakdown. Thus, the system serves as an enzyme to hydrolyze the last phosphate group from ATP. If one measures this hydrolysis in the presence and absence of sodium and potassium, it becomes clear that presence of the

---

6. The suffix *-ase* in an enzyme name usually denotes breakdown of whatever precedes it. An ATPase breaks down ATP by hydrolysis. In this case, energy from the hydrolysis is used to drive the transport of sodium and potassium.

two ions stimulates the reaction. Likewise, one can demonstrate that the addition of ATP stimulates movement of the two ions in the directions mentioned. Thus, ATPase plays an essential role in the function of nerves and, moreover, it is a target of certain pesticides, to which we direct our attention later.

## The Function of Nerve Cells

Nerve cells, which are of central importance in coordinating the activities of most multicellular animals, represent the extreme logical development of a property of all cells, namely excitability. Thus, all cells are excitable insofar as they respond in some way to a stimulus from without, while nerve cells are specifically modified so that the response is their central function. When nerves do respond, the response is primarily of an electrical character, and it is fundamental that the electrical event is transmitted along the length of the elongated nerve cell so that it can serve as an information transfer device within the organism. More simply stated, a nerve is a very long cell, and if one end of it is stimulated an electrical event can be transmitted to its furthermost extensions. Thus, the questions that we might reasonably ask about a nerve cell include those concerning the nature of the electrical event and the description of the mechanism for initiating it.

The external, or plasma, membrane of nerve (and other) cells exhibits an electrical potential difference between interior and exterior of about one-tenth of a volt. Because this potential exists when the cells are not being stimulated, it is called the *resting potential*. However, the term is somewhat misleading: the cell cannot really rest when the potential is there since it is generated by the continuous expenditure of energy. Thus, any interference with respiration or other sources of energy causes it to disappear. The occurrence of an electrical potential implies that there must be an unequal movement of charge across the membrane where it occurs. In the case of nerve cells (and, indeed, all cells) the unequal movement of charge reflects an unequal flux of ions in the following manner: the ATPase causes extrusion of sodium ($Na^+$) and an uptake of potassium ($K^+$) coupled to the utilization of energy through the breakdown of ATP. Because this countermovement of sodium and potassium is equal it does not, in itself, produce a charge difference across the membrane. On the other hand, the required asymmetry of charge movement occurs when the ions return passively in the opposite directions. Thus, when the two ions have been "pumped" in the directions mentioned, they then leak back across the membrane; it turns out that the potassium leaks more readily and faster than the sodium. This asymmetry simply derives from a greater permeability of the membrane for potassium than for sodium. The result is that a rapid flow of potassium *outward* fails to be balanced by a slower *inward* sodium return, so that there is

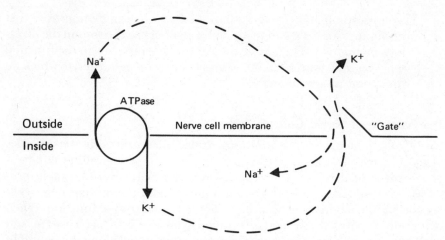

**Figure 3-10.** The action of the sodium-potassium ATPase in the nerve cell. The ATPase causes extrusion of sodium ($Na^+$), and the uptake of potassium ($K^+$), using ATP breakdown as a sort of fuel. When the nerve cell is not excited, the "gate" will be closed and the $Na^+$ and $K^+$ can only leak slowly back across the membrane. Since $K^+$ leaks more rapidly than $Na^+$, there is a net movement of positive charge in an outward direction leading to the resting potential of the membrane (see text). When a nerve is excited, the gate opens (i.e., the membrane suddenly becomes more permeable) and a large flow of positive charge occurs. Since the $Na^+$ influx occurs slightly before the $K^+$ exit, there is a charge imbalance that accounts for the action potential which is the electrical storm that moves rapidly along the nerve and carries its signal. Finally, the original balance of $Na^+$ and $K^+$ is restored through the operation of the ATPase.

a net movement of positive charge outward, with the inner surface of the membrane becoming negative relative to the outside (see Figure 3-10).

When a cell is stimulated by electrical, mechanical, or chemical means, then the pattern of the resting potential is rapidly altered. Thus, if a nerve is stimulated at one end—either artificially or by another nerve—that local region of the cell surface undergoes a change of polarity, with the interior becoming briefly more positive than the outside. This occurs through the "opening" of the membrane to sodium so that the sodium floods inside producing the positive inside surface. This process is rapid, requiring only a few thousandths of a second, and it is followed by an outward potassium movement which tends to restore the original resting polarity of the membrane. Finally, the sodium-potassium ATPase pumps the two ions in opposite directions so that the original resting condition of the nerve is regained. The important part of the process from the standpoint of information transmission is that the charge movements in a localized part of the nerve cell surface induce a similar set of events in the part closest to them, resulting in a wave of these events moving along the nerve.

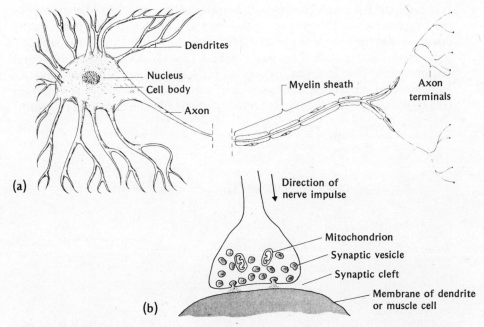

**Figure 3-11 a and b.** The structure of a mammalian nerve cell (neuron). Diagram (a) illustrates the overall structure of a myelinated motor neuron. Diagram (b) represents a large magnification of the terminal part of an axon, showing details of the synapse. (Adapted from Kimball, John W.: *Biology, Third Edition,* 1974, Addison-Wesley, Reading, Mass.)

When two nerves are in contact, there is often the possibility of one stimulating the other across the gap that separates them—the *synapse*. Where this occurs, the message that passes across the gap is in the form of a chemical, termed a *transmitter substance,* the most common example of which is acetylcholine.

$$\begin{array}{c} (CH_3)_3 \\ | \\ N^+ \\ | \\ CH_2 \\ | \\ CH_2 \\ | \\ O \\ | \\ C-CH_3 \\ \| \\ O \end{array}$$

It is essential that such compounds be released on one side of the synapse very rapidly when the pre-synapse nerve is stimulated, and it is equally essential that they be destroyed rapidly after they have caused the stimulation of the post-synapse nerve (see Figure 3-11b). To accomplish this second requirement, post-synaptic nerves stimulated by acetylcholine have an enzyme localized at the synapse called *acetylcholine esterase* that catalyzes cleavage at the molecule to form acetate + choline by the following reaction:

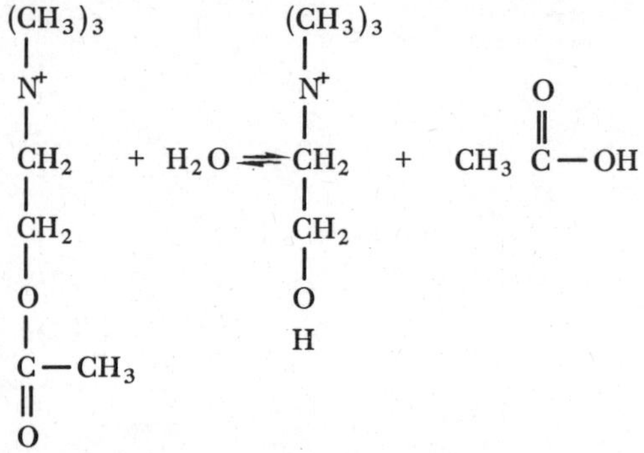

$$\begin{array}{c} (CH_3)_3 \\ | \\ N^+ \\ | \\ CH_2 \\ | \\ CH_2 \\ | \\ O \\ | \\ C-CH_3 \\ \| \\ O \end{array} + H_2O \rightleftharpoons \begin{array}{c} (CH_3)_3 \\ | \\ N^+ \\ | \\ CH_2 \\ | \\ CH_2 \\ | \\ O \\ | \\ H \end{array} + CH_3\overset{O}{\underset{\|}{C}}-OH$$

This enzyme is of interest in the context of this book because it is the site of attack of many organic phosphate insecticides (to be discussed in a later chapter). If a substance interferes with this reaction, the post-synaptic nerve will be stimulated permanently and will be worthless in information transfer. Thus, inhibition of the acetylcholine esterase will lead to complete breakdown of nervous communication, resulting in paralysis and death. This result is desirable from our point of view as long as the nerves are those of the target insect, but unfortunately, human and many other nerves function the same way. We shall see that these insecticides are among the most dangerous ones in current use.

# Chapter 4

# Cellular Protection Against Pollution

Cells and thus organisms possess impressive capabilities directed toward protecting the stability of their internal environment. Cytoplasm remains remarkably constant as to chemical composition, pH (acidity), and physical properties regardless of how much variation occurs in the same measurements in the external medium. In fact, a change in the external environment of cells—great enough to alter substantially the interior of cells—tends to be so drastic that it produces irreversible effects, most commonly, death.

Mankind has devised some rather exotic ways to provide cells with a novel external environment. The portion of a river that is a few miles downstream from the effluent pipes of a large paper mill is remarkable not for its sterility but for the variety of microscopic life prospering there with undeniable vigor.

Moreover, one can think of many examples of components of human pollution that are, in principle, poisonous and yet fail to poison until very high levels are reached. Insecticides are certainly insect poisons and yet are often strangely nontoxic to insects that have been previously exposed to them. Likewise, a number of drugs of the "dope" classification (themselves a sort of self-inflicted pollution), lose their effectiveness in people who have been previously exposed to *them* unless doses are increased. The point of this chapter is that organisms generally, and animals in particular, have internal mechanisms in their cells which serve to protect them against assault from novel compounds in their external environment. We shall now turn to some examples of mechanisms whereby animal cells neutralize the effects of such compounds and maintain their internal constancy, called *homeostasis*, which is so characteristic of life.

## The Membrane Barrier

A compound that fails to penetrate cells is usually no threat to their livelihood. Furthermore, a compound that fails to reach a cell is hardly any threat at all. Most land organisms and many aquatic organisms possess an integument, such as a skin or shell, which serves to protect them against drying out or from mixing their interior with the environment. These integuments also serve to lessen the impact of polluting substances on the interior functions of the organisms. For example, the outer layers of human skin are composed of highly insoluble protein that protects internal organs from both mechanical and chemical injury (see Figure 4-1). Very few chemicals can penetrate human skin; those that do are, with a few exceptions, so corrosive as to simply dissolve the skin and all other tissue. For this reason, compounds that harm humans (and, indeed, those that nurture them) must pass into the interior through one of several weak points in the armor such as the gut or the lungs, both of which are surfaces that exist for the purpose of exchange with the environment.

In general, animals also possess a system for extrusion of harmful material whether it be from external sources or from the waste products of normal metabolic function. For instance, the kidneys of vertebrates and the nephridia of many lower animals can effect the removal of a wide variety of toxic substances. One limitation in this level of protection is that if a substance is toxic enough, it may well harm the system that is engaged in its

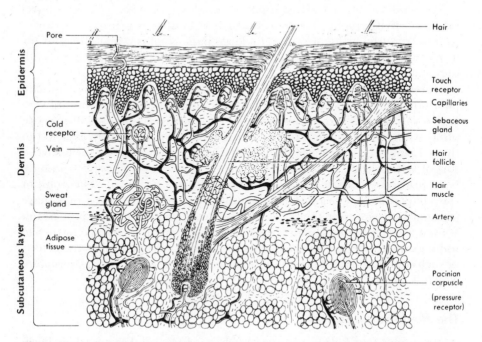

**Figure 4-1.** Section through human skin. Note especially the *epidermis*, which forms an effective barrier against many types of environmental insult. (From Kimball, John W.: *Biology, Third Edition,* 1974, Addison-Wesley, Reading, Mass.)

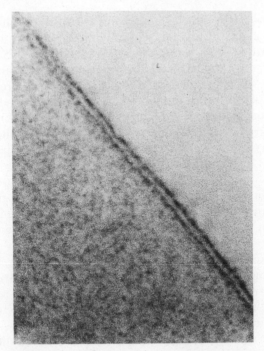

**Figure 4-2.** Electron micrograph of the plasma membrane (magnification is × 250,000). (With permission from J. David Robertson.)

removal. For example, the human kidney is one site of the harmful influences of a number of metallic poisons such as lead and mercury for the obvious reason that these materials tend to be concentrated there on their way to excretion.

In addition to these mechanisms for removal of poisons, and protective coatings such as skin, there is the special integument of cells called the plasma membrane (Figure 4-2). We saw in Chapter 3 that cell membranes are very restricted and specific in their permeability. In other words, they exclude many compounds (and ions) from entry while admitting only the necessary ones. This exclusion is obviously a factor in minimizing the effects of polluting substances. In fact, the requirements of solubility in the design of certain classes of insecticides is instructive. Among the so-called chlorinated hydrocarbons, those compounds that are most soluble in fats are most easily able to penetrate cells as a result of their solubility in the lipid of membranes. These, of which DDT is an example, are among the more potent insecticides simply because the cell-membrane level of defense is most helpless in their presence.

A final type of passive protection against harmful substances in the environment is the ability of several cellular components, notably certain proteins, to bind them. For example, low levels of such metallic poisons as mercury apparently tend to be relatively harmless to humans, and this lack of toxicity may reflect a number of binding sites for mercury on cellular proteins. Thus, if there are a number of sites that bind it readily and binding to these sites does not produce significant harm, a sort of mercury buffer will exist to insure that major harm will not occur until all sites are occupied and the mercury begins to find its way to sites where binding does damage. Such protection would necessarily be temporary as the binding sites would probably be constant in number. Therefore, the buffer would only "buy time" against the poison, freeing the sites for subsequent protection.

The importance of this type of protection is seen in the interesting case of rotenone, an insecticide. Rotenone is a poison because of its effect on cellular respiration. It is equally inhibitory to the respiratory process in mitochondria from most animal (and plant) sources. Specifically, it inhibits respiration to the same degree in insect and mammal mitochondria. On the other hand, the material exhibits a highly specific degree of toxicity when applied to whole organisms. For instance, it is highly toxic (insecticidal) to many insects but not to some others. It kills fish when added to water in low concentrations but remains almost entirely harmless to mammals. Since this poison acts in the same way and to the same degree in the mitochondria of all of these organisms, the difference in toxicity must be due to differences in its ability to *reach* the mitochondria. In other words, the various animals differ in their ability to defend themselves against rotenone owing to different properties associated with their integuments, cellular membranes, or other protective devices.

The most refined case of a system existing for the protection of cells against external materials is that of the immunological defense of warm-blooded vertebrates (and perhaps other animals). This system has evolved because it protects animals against infectious organisms, and it works by "recognizing" the proteins on their surface as foreign. Introduction of such alien protein leads to the synthesis of a second protein able to bind to it in the antigen-antibody complex. The foreign protein is the *antigen* and the protein synthesized in response to it is the *antibody*. This complex eliminates any harmful effects of the protein if it is, itself, toxic, as in the case of a bacterial toxin. Or it limits the invasive capability of a bacterium if the protein is from its surface coat. This special line of defense is mentioned here as an illustration of defense mechanisms in general, although it is of limited applicability to the problem of pollution physiology. Thus, few proteins are important components of pollution (apart from possibly apocryphal and senseless schemes to pollute reservoirs with bacterial toxins during the Second World War). In any case, proteins tend to be little threat owing to their inability to penetrate cellular membranes as a consequence of their large size.

## Cellular Detoxification Mechanisms

Many foreign compounds are modified by cells in such a manner as to render them harmless. Such modification is tantamount to incorporating the compounds into the metabolic web of the cell and to the carrying out of some chemical transformation that either enables the compound to be further metabolized or to be excreted without harm. For instance, if a particular chemical group such as the acidic group, —COOH, is responsible for the toxicity, then blockage or removal of that group is all that is needed to convert the compound to a harmless derivative.

Cells are often able to carry out such transformations on molecules that are entirely novel to them. Thus, many compounds that are products or by-products of industry have only existed on earth for a few years when compared to the time that cells have been in existence. When one discovers that there are mechanisms in cells for detoxification of these compounds, it is tempting to imagine that cells are very foresighted indeed to have anticipated the efforts of the synthetic chemist or, on the other hand, very adaptable to have developed techniques for dealing with the compound when the need arose.

Cells are certainly adaptable, but this sort of phenomenon is hardly a case in point. A more reasonable explanation is seen in the metabolic diversity of cells. In other words, a general rule of metabolism is that cells can convert almost anything to almost any other compound as long as the transformation is chemically feasible. But the fact is that a cell which is contaminated by a particular compound may possess the enzymes required to render it nontoxic. If removal of a particular group is the process required, then that removal is often effected by a set of enzymes which remove the group normally in a quite different context, say, as a part of amino acid metabolism. There is a case in which an enzyme system might seem to exist "for the sake of" the oxidation of drugs and certain other compounds including insecticides. Synthesis of this collection of enzymes from the endoplasmic reticulum is actually stimulated (i.e., the enzymes increased in concentration) by the presence of certain drugs or insecticides that it oxidizes.

Thus, one is again tempted to attribute a form of wisdom to cells to have anticipated the existence of such compounds. This is certainly the incorrect approach from the standpoint of evolution; more reasonably, one should look for a more general function for this set of enzymes and should only feel grateful that the structures of certain drugs and insecticides fall within the limits of specificity of these enzymes, both in their action and in the inductive process that promotes their synthesis.

There are numerous cases of detoxification relevant to the present study, perhaps as many as there are functional groups on the molecules found in pollution. Instead of enumerating them, we shall provide an example which is probably the first such mechanism known. Benzoic acid and related compounds are relatively toxic chemicals which occur in some industrial effluents and also as normal products of certain metabolic processes. It is probably because of the latter that a mechanism for their detoxification occurs in cells. Benzoic acid has the structure

$$\text{C}_6\text{H}_5-\text{COOH}$$

where the —COOH is its *functional group,* which is the group that gives it most of its chemical properties. This compound is made less toxic by formation of an amide linkage with the amino group of the amino acid, glycine.

$$\underset{\text{Benzoic acid}}{\text{C}_6\text{H}_5\text{COOH}} + \underset{\text{Glycine}}{\text{H}_2\text{N}-\underset{\underset{\text{COOH}}{|}}{\overset{\overset{\text{H}_2}{|}}{\text{C}}}} \longrightarrow \underset{\text{Hippuric acid}}{\text{C}_6\text{H}_5-\overset{\overset{\text{O}}{\|}}{\text{C}}-\underset{\underset{\text{COOH}}{|}}{\overset{\overset{\text{H}}{|}}{\text{N}}}-\text{CH}_2} + \text{H}_2\text{O}$$

The reader will note that this reaction is the same as the reaction connecting two amino acids by means of a peptide linkage to form a dipeptide (see Chapter 3). The detoxifying reaction should probably be regarded as a special "use" of a normal reaction of peptide synthesis. This particular mechanism is more general, and other organic acids are also rendered harmless by fusion to other amino acids. For instance, phenylacetic acid, which may be considered as benzoic acid with an extra —$CH_2$— between the acid group and the ring,

$$\text{C}_6\text{H}_5-\text{CH}_2\text{COOH}$$

is coupled to the amino acid, glutamine, to form the following derivative:

$$\text{C}_6\text{H}_5-\text{CH}_2\overset{\overset{\text{O}}{\|}}{\text{C}}-\overset{\overset{\text{H}}{|}}{\text{N}}-\underset{\underset{\underset{\underset{\underset{\underset{\text{O}}{\|}}{\text{C}-\text{NH}_2}}{|}}{\text{C}-\text{H}_2}}{|}}{\overset{\overset{\text{COOH}}{|}}{\text{C}}}-\text{H}$$

Finally, phenol, which is also acidic and is often an important component of effluent from various industries,

$$\text{C}_6\text{H}_5\text{—OH}$$

is altered to form phenylsulfuric acid which is, despite the name, relatively harmless to the cell.

$$\text{C}_6\text{H}_5\text{—OSO}_2\text{OH}$$

In all such cases, the derivative formed by the "detoxifying" reaction is then excreted in the urine.

## A System for the Oxidation of Drugs and Pesticides

Reactions leading to lowered toxicity of foreign compounds often have oxidations as their basis. For instance, alcohols are often oxidized to form the corresponding acids; the generalized reaction follows, with "R" representing an unspecified part of the molecule.

$$\text{R—CH}_2\text{OH} \xrightarrow{[\text{O}]} \text{R—COOH}$$

It is stated in Chapter 3 that oxidations in cells are often carried out by systems including the iron-heme proteins called cytochromes, and these types of detoxifying oxidations are not exceptions. Thus, such reactions are carried out by systems of oxidases that contain a cytochrome as well as flavoproteins and other heme proteins and which occur in the endoplasmic reticulum fractions of cells. Preparations derived from endoplasmic reticulum are often called *microsomes*. Thus, the enzyme system under discussion is often referred to as the *microsomal oxidase*. These enzyme systems are also called mixed function oxidases since they both oxidize a reduced coenzyme (such as NADH) and oxidize another substrate molecule, often by means of a hydroxylation (see Figure 4-4). Expressed differently, these oxidases reduce the diatomic oxygen molecule ($O_2$) with electrons from the two different sources (the coenzyme and the substrate), and one of the oxygen atoms eventually becomes part of a hydroxyl (—OH) group on the substrate.

Without considering their mechanism, these enzyme systems are important in oxidizing many compounds found in pollution, including insecticides. However, the primary action of these systems—the action for which

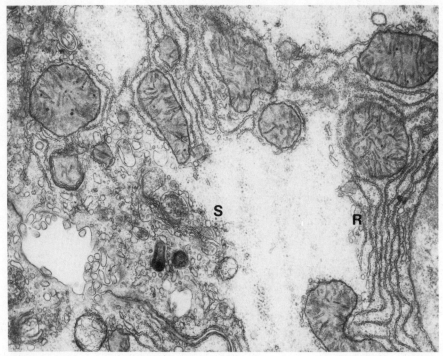

**Figure 4-3.** Electron micrograph of mouse liver (magnification is × 21,000). Regions of rough (R) and smooth (S) endoplasmic reticulum are denoted. (Micrograph by M.S.C. Birbeck, in E.J. Ambrose and Dorothy Easty, *Cell Biology*, Addison-Wesley, 1971. Used with permission from Thomas Nelson & Sons, Ltd., London.)

they originally evolved—was probably the oxidation (or hydroxylation, a different but related process) of steroid compounds, a class of chemicals that includes many of the sex hormones as well as cholesterol. These enzyme systems function with other compounds such as DDT by virtue of a relatively low degree of specificity combined with some similarity between the different classes of compounds in question. This low specificity greatly extends the function of the enzymes as new polluting compounds enter the human system.

An interesting aspect of the endoplasmic reticulum (microsomal) oxidase systems is the fact that their formation is induced by some of the compounds upon which they act. In other words, they are enzyme systems whose synthesis is stimulated by the presence in cells of their substrates. Not only is their synthesis stimulated by their natural substrates, but also by the more recent "unnatural" substrates such as the insecticides mentioned. Thus, cells of animals that have been fed DDT will exhibit a higher activity of these oxidases including a considerably higher content of the cytochromes such as $b_5$.

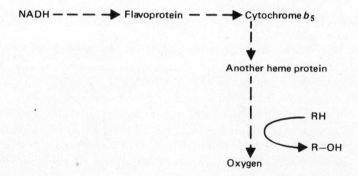

**Figure 4-4.** An oxidation system from the endoplasmic reticulum. Dashed arrows denote movement of electron; RH is the organic molecule being hydroxylated. Since preparations made from disrupted endoplasmic reticulum are often called microsomes, this system is often referred to as a *microsomal oxidase*.

An important aspect of the induced synthesis of these enzymes is the fact that the enzyme does not "know" what compound stimulated its synthesis. This means that one inducing compound will produce an increase in the microsomal system which will then be more active with other compounds as well. Since an increased activity of the system represents an adaptation to the presence of whatever induced it, the adaptation can be said to cross over and provide benefits to the organism in the form of increased ability to cope with other compounds as well.

To place this effect in a more specific context, it is well known that habitual ingestion of alcohol (ethanol) by humans leads to its increased rate of metabolism and probably a consequential increased tolerance to it. It is interesting that the increased metabolism extends to other drugs such as pentobarbital, a sedative; antipyrine, a fever-reducing drug; and tolbutamide, a drug used to help control diabetes. This observation suggests that it is extremely important for a physician to know that he is treating an alcoholic when he prescribes any of these drugs, so that the dosage can be correctly determined. The situation is actually somewhat more complicated since *large* doses of alcohol produce the reverse effect, namely an inhibition of the microsomal system. This inhibition accounts for the well-known increase in the effects of certain drugs such as barbiturates, including pentobarbital, when people "wash them down" with large amounts of alcohol.

Similar crossover effects occur in the case of mammalian metabolism of insecticides. For instance, treatment of rats with low levels of chlordane decreases their sensitivity and increases their metabolism of other insecticides including DDT, heptachlor, aldrin, and dieldrin. In the case of humans, a similar effect occurs. Moreover, it has an implication of great significance in human health that relates, again, to the low specificity of the microsomal oxidases.

Thus, people who are exposed to large amounts of DDT or other insecticides are able to metabolize the drug antipyrine more rapidly than can unexposed subjects. Moreover, they also exhibit an elevated ability to oxidize a variety of steroids including several that are important drugs and several more that are normal body constituents, namely, steroid hormones.

It also appears that exposure to either of these classes of insecticides or drugs provides an increased detoxification of a third class of compounds which are *carcinogenic,* i.e., tumor-producing agents. Increased levels of insecticide in humans can be expected to produce a number of rather different results through their inductive action on the microsomal oxidase system of enzymes. Some of the effects are favorable, such as the decreased sensitivity to the insecticides themselves, or to carcinogenic compounds. Other effects produce changes in the hormonal balance of the person, possibly with profound deleterious effects. Still others lead to changes in drug metabolism that must be considered if such drugs are to be used for therapeutic reasons.

An additional remark should be made regarding the effect of such environmental chemicals as insecticides on steroid metabolism. We will see in a later section that at least two of the important effects of insecticides on warm-blooded animals may occur through changes in hormone balance. Thus, insecticide-produced sterility and other sexual problems in rodents, such as altered breeding behavior, may occur through increased oxidation of sex hormones which are steroids. In addition, the thin eggshell syndrome in birds, which has led to drastic reduction in the numbers of certain species, may occur due to changes in the steroid-mediated control of calcium metabolism and, hence, eggshell deposition.

Thus, several vitamins have structures that permit them to act as substrates for the microsomal oxidase. For example, vitamin D is the steroid

Table 4-1. Some examples of compounds that influence or are influenced by the microsomal oxidase system.

---

**Compounds that are oxidized**
DDT, lindane, chlordane, aldrin, quinine, warfarin, pentobarbital, antipyrine, vitamin D, testosterone

**Compounds that induce synthesis of oxidase enzymes**
phenobarbital, antipyrine, nicotine, DDT, lindane, chlordane, aldrin, ethanol

**Compounds that inhibit the oxidase enzymes**
piperonyl butoxide, carbon tetrachloride

vitamin that functions in the regulation of calcium metabolism. Increased oxidation due to the effect of DDT may lead to a vitamin deficiency even though there would normally be an adequate intake of the compound. Similarly, vitamin K is involved with the blood-clotting process and perhaps with other cellular events. It too is oxidized by microsomal enzymes so that it may also become limited when these enzymes are elevated due to the presence of environmental chemicals or drugs. Thus, excessive bleeding due to vitamin K deficiency has occurred in infants whose mothers were taking certain drugs over an extended period of time. There is no reason why similar effects could not occur after long exposure to other inducers of the oxidation system, such as insecticides.

Finally, since many insects possess microsomal oxidase systems whose synthesis, like the mammalian ones, is induced by insecticides, the pesticide industry has devised a number of *synergists* that are included with insecticides and which act to inhibit the insect's oxidase system. A good example of such a pesticide synergist is the chemical piperonyl butoxide

$$\text{structure: benzene ring fused with a methylenedioxy (-O-CH}_2\text{-O-) group, with substituents } CH_2OCH_2CH_2OCH_2CH_2OC_4H_9 \text{ and } CH_2CH_2CH_3$$

which renders insects less able to oxidize (and, hence, detoxify) insecticides. Unfortunately, this compound also inhibits human microsomal oxidations and may interfere with our detoxification processes in a similar manner. Luckily, existing evidence suggests that the compound must be present in much higher amounts in mammalian tissue as compared to insects to have a comparable effect and so does not appear to be a major threat.

# Part II

## Cellular Aspects of Pollution

# Chapter 5

# Physiological Effects of Insecticides

Pesticides are chemicals whose utility resides in their ability to kill certain organisms regarded by humans as undesirable for one reason or another. Thus they are by definition toxic and all other conditions being equal, the more toxic they are, the more highly regarded they become by their users. In addition, for a pesticide to be useful it must be highly specific: it must kill or deter the target organism while remaining relatively harmless to other organisms. For example, an insecticide that shows equal toxicity for mosquitoes and for the human holding the spray can is not likely to become a commercial success. To cite a less extreme example, insect poisons that prevent reproduction in birds are presently declining in popularity, if not use, even though one could argue that they are relatively safe and specific, being much more toxic to insects than to hawks or pelicans.

This chapter is concerned with the manner in which insecticides kill. It considers both their action upon target organisms and on what one might term innocent bystanders. Since it treats these matters almost entirely on the cellular level, the reader is urged to return when necessary to Chapter 3 where cellular physiology is introduced. Also, the numerous organic chemical structures that are necessarily included in this chapter may be examined with the aid of the Appendix.

## Determination of Toxicity

Since we will consider the toxic effects of certain compounds on a variety of organisms, we must first discuss toxicity in a general sense. In the first place, it is usually necessary to know how toxic a given substance really is. This is not necessarily a simple matter; one needs, for example, to be precise as to the measure of toxicity and as to the nature and size of the organism on which the substance is to be used.

Because the ultimate effect of a toxic material must be the death of an organism, the most widely used measure of toxicity relates to the ability of a substance to kill. This measure is the $LD_{50}$, which is the amount of a chemical required to kill 50% of a test population of the organism studied. This measure, in turn, raises the question of how that amount should be expressed. First, one should consider the size of the organisms being killed. In general, the larger an organism is, the more of a particular chemical is required to have an influence on it. It is intuitively obvious that a dose of poison just sufficient to kill a mouse will be unlikely to topple an elephant (unless, of course, one has discovered a chemical that exploits some special difference between the physiological properties of mice and elephants).

If one is to test the action of a substance on members of a single species, one would generally expect the smaller individuals to be the most sensitive. For this reason, a rational measure of toxicity should be expressed in terms of the mass of the organism, which means that the $LD_{50}$ should be given as amount per gram, or kilogram, of organism. Finally, the amount of the chemical should be expressed in terms which consider its molecular weight, as it is the number of molecules of a chemical that most closely relate to its action rather than the number of grams. Again, all other conditions being equal, the larger the molecular weight of a compound, the more weight of it that must be added to produce an effect. Thus, the most rational units for the $LD_{50}$ become moles (or millimoles or micromoles) per gram (or kilogram) of organism.[1]

This last point is often overlooked and the most common form assumed by the $LD_{50}$, as it appears in scientific literature, is on a simple weight basis, that is, mg per kg. An indication of the sort of values for $LD_{50}$ encountered in the case of insecticides is given by Table 5-1 which shows the toxicity of several insecticides with different organisms. Note both the small amounts required for insecticidal action and the high degree of selectivity. Reflect also on the idea of a kilogram (slightly over two pounds) of mosquitoes.

It is important to realize that the means of administering a poison influences its toxicity. In Table 5-1, the insecticides were administered to mammals orally and to insects topically (on their surface). Had the insecticides been applied to the mammals by the topical method, the toxicity would have been considerably less.

A final aspect of toxicity is related to the use of *coinsecticides,* or synergists. These are compounds, not necessarily toxic themselves, which render other compounds more toxic than when used alone. For example, pyrethrins are extremely potent insecticides in their own right, but when they are applied with any one of several synergists isolated from sesame oil, they became dramatically more potent. It is interesting that compounds showing activity as

---

1. A mole of a substance is its molecular weight in grams. Molecular weights, in turn, are the sums of the atomic weights of the constituent atoms. Atomic weights are expressions of the mass of the individual atoms of the different elements as relate to oxygen, which is considered to have an atomic weight of 16 daltons.

Table 5-1. Toxicity of various insecticides.

| Compound | Organism | $LD_{50}$ (mg/kg) |
| --- | --- | --- |
| DDT | bee | 114 |
| DDT | cockroach | 10 |
| DDT | housefly | 8 |
| DDT | rabbit | 300 |
| Dieldrin | housefly | 1.3 |
| Dieldrin | rat | 87 |
| Pyrethrin | Japanese beetle | 40 |
| Pyrethrin | mosquito | 1 |
| Pyrethrin | guinea pig | 1500 |

Source: R. D. O'Brien, *Insecticides—Action and Metabolism* (New York: Academic Press, 1967).

synergists with pyrethrins also enhance the toxicity of some chemically unrelated insecticides.

## The Classes of Insecticides

Throughout the remainder of this chapter it will become evident that (1) different insecticides have different effects on insects and other organisms, and (2) there are a great number of different insecticides. For these reasons we are dividing the insecticides into six main categories and then examining each with regard to its effect on insects and then on nontarget organisms. In some respects the classification is a bit artificial, but it does promote clarity. These six main classes of insecticides are shown in Table 5-2 with chemical structures of some examples. The column labeled "remarks" provides the reader with several attributes to associate with each class. Even if one has only minimal understanding of organic chemistry, a careful examination of these structures will provide an additional association to aid in keeping material organized. If one is not primarily a chemist he should not feel self-conscious in picturing DDT as the compound with the two rings and all the chlorines (Cl) hanging off of it or chlordane as possessing the funny-looking ring with the line through it.

Another property of insecticides mentioned in Table 5-2 is the *persistence* of a compound. This term refers to the relative length of time that a

Table 5-2. Classes of insecticides.

| Class | Example | Remarks |
|---|---|---|
| 1. Organo-chlorine insecticides | DDT $Cl-C_6H_4-CH(CCl_3)-C_6H_4-Cl$ | (a) used since 1940<br>(b) kill insects by action on nerves<br>(c) very persistent |
| 2. Organic phosphates | Parathion $(C_2H_5O)_2P(S)O-C_6H_4-NO_2$ | (a) related to nerve gas<br>(b) low persistence<br>(c) inhibit esterase |
| 3. Cyclodienes | Chlordane (chlorinated cyclic structure) | (a) also act on nerves<br>(b) used since 1950 |
| 4. Carbamates | Sevin $OC(O)NHCH_3$ on naphthalene | (a) like organic phosphates in inhibiting acetylcholine esterase<br>(b) only effective against some insects |
| 5. Pyrethrins | Pyrethrin * | (a) some are naturally occurring<br>(b) act on nerves<br>(c) low persistence |
| 6. Rotenoids | Rotenone | (a) naturally occurring<br>(b) act on cell respiration<br>(c) low persistence |

*The structure is that of pyrethrolone which is coupled to an acid, such as pyrethric acid, to form the insecticidal pyrethrin.

compound remains in the environment after being deposited by humans. Persistence is inversely proportional to the ease with which organisms (chiefly microorganisms) can metabolize a given compound. In general, the more bizarre a compound's structure is—i.e., the more dissimilar its structure is to compounds found in the natural world—the more unlikely that bacteria possess enzymes which will degrade it. Since insecticides are synthesized and selected for their ability, among other attributes, to resist degradation by the target organisms, they are frequently highly resistant to bacterial degradation as well. Their effectiveness partly resides in their persistence. In any case, since insecticides in the first two categories are products of synthetic chemists, they are generally much more persistent than the remaining categories. The rotenoids possess the virtue of eventually being destroyed by light so that they have a most convenient self-destruct aspect. Since the rotenoids and the pyrethrins are naturally occurring compounds, they have short half-lives in the environment due to the presence of bacteria able to break them down.

## How Insecticides Kill Insects

Since the author of this book does not happen to be an insect and since the book is chiefly concerned with the harmful (from the human viewpoint) effects of polluting substances, the present chapter must be concerned primarily with the influence of insecticides on nontarget organisms. However, there appear to be enough similarities between the manner in which insecticides interfere with the physiology of insects and the manner in which they harm other organisms for us to carefully examine the destruction of insects.

As a general rule we can assert that the insecticides in the categories listed in Table 5-2 kill insects by fouling up the insect's nervous system. As we shall

Table 5-3. Persistence of some insecticides.

| Class of insecticides | Half-life in environment |
|---|---|
| Organo-chlorines | a few years |
| Organic phosphates | a few weeks |
| Carbamates | about a week |
| Pyrethrins and rotenoids | a few days |

see, the several classes accomplish this in different ways. Moreover, in no case is our knowledge so complete as to allow us to believe that we understand the entire process from ingestion of the compound by the insect to the irreversible disorder leading to death. For example, we may know that a certain compound is a nerve poison without really knowing why poisoning an insect's nervous system leads to its demise. Finally, the assertion that insecticides produce their effect by attacking the insect nerve may only reflect our having examined nerves most carefully. If one studies the insect nerve to the exclusion of all other organs and systems, the effect of a flyswatter will be termed neurotoxic, although the insect be squashed flat. The original assertion is probably valid, but one is cautioned to be wary.

## The Effect of DDT

There is considerable evidence that the insecticidal effects of DDT are at the level of the insect's central nervous system. Indeed, as we shall see later, the harmful effects on animals other than insects may also occur at that locus. One of the first symptoms of DDT intoxication[2] in insects is often an excessive activity associated with tremor as if the nervous system were being abnormally stimulated. Later these symptoms disappear and are replaced with decreasing nervous function until weakness, paralysis, or death occurs. Finally, when a sensitive insect is dissected and DDT is applied to separate organs or tissues, it appears that only nervous tissue is sensitive to very low concentrations of the poison. All of these features of DDT toxicity lead to the same conclusion: DDT is able to poison the nervous system of insects.

With this evidence, one must admit that many of the details of the action of DDT on nerve cells elude us, and that the causal connection between effects on the nervous system and the killing of the insects remains to be ascertained. Several important features of its action on the nervous system have, however, been known for some time. First, it seems clear that DDT interferes with the process of transmission of an impulse along a nerve cell rather than the alternative possibility of influencing synaptic transmission between separate cells. Moreover, the lowest concentrations of DDT act on sensory but not motor nerves. In studying isolated nerves, one can probably observe the basis for the excessive nervous activity associated with DDT intoxication in the form of a sort of impulse multiplication: when a single impulse travels along a nerve into a region where DDT has been placed on the nerve, then the impulse becomes a volley of impulses lasting a considerable time. Additionally, there is evidence that DDT alters the "shape" of the action potential, which is its magnitude in relation to time. Without

---

2. In biological terms, intoxication refers to any toxic effect and, in contrast to ordinary usage, does not identify the particular poison used.

describing this final alteration in detail, it may be stated that the character of the alteration is such that it could be explained by postulating an altered ability of the nerve cell membrane to admit sodium or potassium. Therefore, it is of great interest to learn that addition of DDT to insect (cockroach) nerves produces an increased permeability to radioactive potassium.

One is led to the point of believing that the action of DDT is the result of an interaction with the nerve cell membrane such that its permeability properties and therefore its excitability properties become altered. This theory is attractive in its ability to explain in principle most features of DDT toxicity. It is also entirely plausible in view of the high solubility of DDT in lipids (fats), such as those comprising the membrane of nerve (and other) cells. The only difficulty with this type of explanation is that an important link in the logic is missing: we are stymied as to why DDT is able to produce significant changes in ion permeability in nerve membranes. Needless to say, this point is under active consideration in many laboratories. We are also uncertain as to the exact connection between increased activity in *sensory* nerves and the death of the insect, although one theory proposes that the insect succumbs to a sort of terminal exhaustion due to the extremely high level of stimulation.

If there are important features of DDT action yet to be understood, our knowledge of the action of most other insecticides is quite rudimentary. Lindane, another chlorinated hydrocarbon, is believed to act in a manner similar to DDT, largely on the basis of its similar effects on intact insects. The class of cyclodienes which include the widely used insecticides, chlordane, aldrin and dieldrin, produce some DDT-like effects on isolated nerves causing, for example, volleys of impulses. On the other hand, evidence reveals that their influence on the intact insect resides in their interacting with the ganglia comprising their central nervous system. It is clear in these cases that the target is at some part of the nervous apparatus of the insect, and evidence points to interaction with the cell membrane of the nerves as in the case of DDT.

## Organic Phosphates

In the case of the organic phosphate insecticides, the action is clearly associated with the nervous system but apparently on a different basis from that of DDT. It is regrettable that the early impetus for the development of organic phosphate technology was military: these compounds include the exceedingly toxic (to humans) "nerve gases" developed in a number of countries during and subsequent to the Second World War. The use of similar compounds as insecticides has been widespread for some time. It has become apparent that they are, in one sense, "good" insecticides in being readily degraded by microorganisms so that they do not reach high levels in

# Physiological Effects of Insecticides

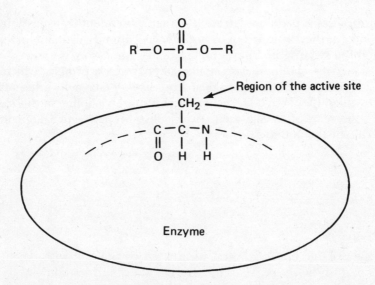

**Figure 5-1.** An organic phosphate compound bound to the serine hydroxyl group of acetylcholine esterase.

similarity to military nerve gas by their high levels of toxicity. A number of gruesome accidents have occurred involving humans coming into contact with concentrated solutions of organic phosphates which led to severe illness and death.

Any aspect of technology with clear military overtones becomes the object of active study, and our knowledge of the organic phosphates and their effects on organisms is quite extensive. There is massive evidence that the primary insecticidal (and homicidal) effects of such compounds reside in their property of inhibiting the enzyme, cholinesterase. It is recalled from Chapter 3 that this enzyme plays a vital role in nerve transmission at many synapses by eliminating acetylcholine after it has been released and has served to fire off the subsequent nerve cell beyond the synapse. The enzyme does this by hydrolyzing the acetylcholine to form acetate plus choline. In other words, it breaks a bond by addition of $H_2O$ "across" it.

$$CH_3COOCH_2CH_2\overset{+}{N}(CH_3)_3 \rightleftharpoons CH_3COO^-$$

$$+ H_2O \qquad\qquad + HOCH_2CH_2\overset{+}{N}(CH_3)_3$$

If this process is prevented, it is clear that subsequent transmission will be prevented as the continued presence of acetylcholine will maintain the nerve beyond the synapse in a perpetual state of stimulation.

The enzyme cholinesterase is, like all enzymes, a protein, which means that it is composed of amino acids. And, like all enzymes, it has a specific region called the "active site" where it binds (acetylcholine in this case) and where the reaction is catalyzed. In this instance the active site contains a molecule of the amino acid, serine,

$$+NH_3-\underset{COO^-}{\overset{CH_2OH}{\underset{|}{\overset{|}{C}}}}-H$$

and the organic phosphate insecticides act by binding with the hydroxyl (OH) part of the serine molecule. In fact, they bind so tightly as to form a new bond between themselves and the serine so that they, in effect, become a fairly permanent part of the now inactivated enzyme molecule. Thus, the active site of the cholinesterase becomes blocked and, as we have seen, nerve transmission is likewise blocked. In the case of mammals, blockage of nerves or synapses leads to extremely adverse situations owing to cessation of function of breathing muscles. In the case of insects the situation is less clear: as we noted in the case of DDT, the causal chain from nerve impairment to death is not yet fully understood.

Many organic phosphates are harmless until they are *activated* within the organism. They essentially require that the organism dig its own metabolic grave by carrying out, with its own enzymes, the final reaction required to render the compound sufficiently toxic. The reaction in question is frequently the replacing of a sulfur adjacent to the phosphorous atom with an oxygen with, for instance, parathion

$$\underset{CH_3CH_2O}{\overset{CH_3CH_2O}{>}}P(=S)-O-\underset{}{\bigcirc}-NO_2$$

being altered to form paraoxon.

$$\underset{CH_3CH_2O}{\overset{CH_3CH_2O}{>}}P(=O)-O-\underset{}{\bigcirc}-NO_2$$

## Naturally Occurring Insecticides

*Rotenone.* One of the oldest (in terms of use) and nonetheless most effective insecticides is rotenone whose structure is shown in Table 5-2. This material is a naturally occurring substance which is obtained from certain tropical plants and has been used by primitive people for millennia as a fish poison. Rotenone is still employed for that purpose such as when ponds are "reclaimed"—i.e., when mixed populations of "trash fish" are eliminated prior to the stocking of waters with desirable fish. Rotenone is well suited for this purpose because it is spectacularly degradable, both by microorganisms and by the photochemical action of sunlight. Thus, after the poison is used in a pond, one need only wait a few days before stocking the new fish with impunity. For over a century it has been known that rotenone is an effective poison for many insect species as well, and it is widely used for this purpose.

There is evidence that the action of rotenone on fish and on insects is essentially by means of the same mechanism and that, in contrast with the insecticides considered to this point, this mechanism is not associated with a specific blockage of nerve transmission. Rather, it appears that rotenone is an effective inhibitor of cellular respiration acting in a manner analogous with that of cyanide and carbon monoxide. The reader should recall from Chapter 3 that energy production in animal cells occurs chiefly through oxidation of foodstuff molecules by means of a chain of enzymes called the respiratory chain. The passage of electrons down the chain toward oxygen is, in a manner not yet completely understood, coupled to the synthesis of adenosine triphosphate (ATP). ATP is, in turn, the energy source[3] for practically all events in a cell which require energy, such as muscular contraction, nerve conduction, active transport across membranes, and many chemical syntheses. In the absence of ATP no active cell can survive for more than a few seconds. This constant and strict requirement for ATP accounts for the rapid death that ensues when animals are deprived of oxygen or when their respiratory chain is blocked by cyanide.

While cyanide and carbon monoxide inhibit the respiratory chain near the oxygen end, rotenone inhibits at the end near the foodstuff (substrate) molecule, specifically between NADH and a flavoprotein molecule (see Figure 3-8). "Between" these two electron carriers means that addition of rotenone to cells causes a damming of the chain at that point: electrons can move only as far as NADH but not as far as the flavoprotein. Expressed in different terms, NADH becomes more reduced and the flavoprotein

---

3. The energy is "transferred" by the coupling of ATP hydrolysis

$$ATP + H_2O \rightleftharpoons ADP + phosphate$$

with the energy-requiring process.

remains oxidized.[4] This cessation of electron flow down the chain stops ATP synthesis thereby killing the cells and the organism.

The reason rotenone kills insects and fish effectively while being relatively safe for most warm-blooded animals requires comment. The fact is that rotenone is a potent inhibitor of the respiratory chain even in mammals. For example, in mitochondria (the locus of the respiratory chain) isolated from rat liver cells, rotenone inhibits NADH oxidation completely at a concentration of about 1 micromolar (or $10^{-6}$ molar). This means that even there it is a potent inhibitor of the respiratory chain, being inhibitory at about 1 one-hundredth of the concentration of cyanide required to poison respiration. However, an intact rat is quite insensitive to rotenone even when it is given orally with the $LD_{50}$ at about 130 mg/kg. This suggests two possibilities. First, rats may metabolize rotenone transforming it into a harmless compound, that is, they may have enzymes that inactivate it. Another possible reason for low toxicity is the inability of the poison to reach the site within the mitochondrion where inhibition takes place. The molecule might be effectively destroyed under the acid conditions of the stomach (which it probably is), or it might fail to penetrate lining of the digestive system to enter the blood. Once in the blood where the poison could be communicated around the body, it could also become bound to serum proteins and inactivated. There is at least some evidence that this is also possible. It seems obvious that important and subtle differences exist between different organisms with regard to inactivation or transport of rotenone, since enormous differences in sensitivity occur even in closely related animals. For instance, pigs are much more sensitive than other mammals while, among the insects, the Japanese beetle is about a hundred times more sensitive than the American cockroach.

Much effort has been expended in trying to learn the mechanism whereby rotenone inhibits the respiratory chain, largely because rotenone is widely used as an inhibitor in order to study the chain. A detailed discussion of this topic is beyond the scope of this book, but one comment may be instructive. As a general rule, when one is trying to ascertain why a particular compound interferes with a process, it is enlightening to look for structural similarities between the inhibitor and components of the system being inhibited. Recall from the preceding discussion that enzymes have active sites where they bind the substances with which they react. These sites have a special geometry which bears a sort of lock-and-key relationship with the shape of the molecule with which they react. Often a second molecule with similar

---

4. This is because reduction means "addition" of electrons to a molecule. Oxidation means removal of electrons. In the respiratory chain the substrate molecule is oxidized (finally by oxygen) since electrons are withdrawn from it and passed along to oxygen.

geometry will block the site and therefore interfere with the proper reaction of the enzyme. An excellent example of such competition for an active site is the inhibition of the oxidation by the respiratory chain of succinate

$$\begin{array}{c} COO^- \\ | \\ CH_2 \\ | \\ CH_2 \\ | \\ COO^- \end{array}$$

by the similar molecule, malonate, which has the following similar structure:

$$\begin{array}{c} COO^- \\ | \\ CH_2 \\ | \\ COO^- \end{array}$$

This type of molecular imitation often underlies inhibition of biological systems. As far as the action of rotenone is concerned, a useful approach to understanding its action may be to compare its structure

to that of the ring system which forms the active region of the flavoproteins of the respiratory chain.[5]

---

5. This ring system is called the isoalloxazine ring. It is shown in the oxidized state. When reduced (by addition of electrons) two hydrogens are added to it as indicated by the arrows.

The possible similarities between the B,C,D rings of rotenone and those of the flavoprotein suggest that the principle of molecular imitation may be in operation here and that the poison may be blocking a site where the flavoprotein ring must bind to accept electrons from NADH.

Moreover, cyanide, which we saw inhibits the respiratory chain at the oxygen terminus, is often used as an insecticidal fumigant. This is either applied as the gas, HCN, or it is sprayed as an organic thiocyanate compound, R—SCN, which is relatively nontoxic but is converted into cyanide in the cells of any target insect. Finally, compounds which "uncouple" ATP synthesis from cell respiration (and which, thus, prevent ATP formation) are used as insecticides. For example, organic tin compounds are members of this class of agents and are potent insecticides used against mites.

*Pyrethrins.* These highly selective and nonpersistent insecticides were originally isolated from members of the daisy family. They have been used commercially for over a hundred years and have long been the basis for a vigorous flower-growing industry. In a time of increasing ecological awareness, these insecticides are enjoying a resurgence of favor owing to their very low toxicity for nontarget organisms.

It is certain that these compounds kill insects by an effect on their nerves, but the nature of that effect is by no means understood. Pyrethrins have the interesting property (and the virtue) of acting very rapidly to cause practically instant paralysis of the insect while many other insecticides require several hours. Study of individual insect nerves indicates that these compounds produce a number of changes in (1) the shape of the action potential, (2) the negative after potential, and that (3) under some conditions they lead to volley impulses from a single stimulus. The molecular or ionic basis for these changes remains to be explained.

*Nicotine.* This insecticide is a plant natural product and is used as a nerve poison for target and nontarget organisms. Soon after the tobacco plant was discovered in the New World, extracts from its leaves were used as insecticides. Nicotine presents the advantage of being naturally occurring and therefore readily biodegradable. It also presents the linked disadvantages that it is considerably toxic for vertebrates (including humans) and is required at relatively high concentrations to kill insects. In the case of vertebrates, nicotine acts upon nerve junctions where acetylcholine is involved in transmission. It is believed that it acts by mimicking acetylcholine, binding to the acetylcholine receptor. The reader will note that this is not a simple case of molecular imitation, as nicotine and acetylcholine

$$\begin{array}{c} CH_3 \\ | \\ CH_3 - N^+ - CH_3 \\ | \\ CH_2 \\ | \\ CH_2 \\ | \\ O \\ | \\ O = C - CH_3 \end{array}$$

have relatively little in common as far as structure is concerned; however, in both cases the configuration around the nitrogen atom appears to be of great importance. In the case of insects, a similar mechanism seems to be in operation with the poison acting upon the central nervous system where transmission appears to involve acetylcholine. Although not as rapidly effective an insecticide as a pyrethrin, nicotine does cause paralysis in a few minutes when applied in a sufficient dose—an effect supporting the view that it is a nerve poison.

## Harmful Effects of Insecticides on Nontarget Organisms

It has been demonstrated in the previous section that in many cases insecticides have the same effects on target and nontarget organisms; it merely requires more to accomplish the same harm in the nontarget case. In this section we shall enumerate some examples where insecticides produce seemingly unrelated and sometimes surprising effects on organisms that humanity had not really intended to harm. In some instances it may be possible to argue that the effects are not really so unrelated. For instance, it will be seen that certain insecticides cause adverse effects in membrane transport in some marine organisms. Although it appears that such effects are not to be described as nerve poisoning, it must be remembered that the movement of ions across membranes is the very basis of nerve transmission. In other cases, the high toxicity of insecticides for insects prevents other analogies from being drawn. For instance, if DDT produces sterility in a rodent, we shall never know if it has the same effect on an insect that is killed rapidly at much lower concentrations than those required to produce the sterility effect.

Much of the controversy about pesticides has revolved around DDT because of its wide use and relative persistence in the environment. It is very soluble in fats and very insoluble in water so that, on a worldwide basis, the enormous amounts injected into the environment tend to become concentrated in the fat deposits of animals and humans and in such other fat-rich substances as the membranes of cells and fat droplets of milk.

The harm that DDT does to nontarget organisms accurately reflects a close association with cellular membranes; in probably all cases, a feature of the harm relates to transport of molecules or ions[6] across such membranes. It will be recalled that DDT interferes with nerve transmission in insects because of a probable action on the movement of potassium or other ions across the nerve cell membrane. Higher concentrations of DDT on mammals produce similar effects so one must begin with nerve toxicity in any list of harm that DDT or other "chlorinated hydrocarbons" does to nontarget animals. Extremely high levels of these compounds must be introduced to produce such effects, and any significant harm produced on a worldwide scale must necessarily be more subtle.

There is increasing evidence that such subtle effects do occur and are of concern in their relation to the fertility of animals. For example, chlordane has been shown to decrease fertility in rats, while aldrin disturbs their estrus cycle and thereby interferes with breeding. These alterations are believed to be related to the demonstrated interference of a variety of chlorinated insecticides with metabolism of steroids, including the sex hormones in the regulation of the internal chemistry and behavior of animals, both of which are extremely complicated. However, it can be simply stated that any change in relative amounts of hormones produced by an abnormal level of breakdown promoted by insecticides cannot fail to cause profound changes in the physiology or activities of an organism. Thus, changes in fertility after administration of such compounds should not be surprising, nor should it be unexpected that administration of DDT to newborn female rats leads to a constellation of alterations. For instance, these rats reach breeding condition at an earlier age than controls do. In addition, as the rats become older they are progressively unable to produce egg cells—an inability that is reflected in degenerative changes in their ovaries, as viewed with a microscope.

It should not be concluded that DDT and related compounds confine their nontarget effects to the animal kingdom. DDT analogs have been observed to be toxic to both terrestrial and marine plants. In some strains of barley, for example, it appears that the compounds interfere with photosynthesis in chloroplasts probably due to inhibition of light-induced electron transfer in the photosynthetic cytochrome chain.

### Effects of Insecticides on Birds

Changes in the breeding capability of mammals noted in the previous section are mirrored in the case of birds. In both cases, the basis for harm

---

6. An ion may be regarded simply as an atom or molecule that bears an electrical charge.

Physiological Effects of Insecticides 79

(a) (b)

**Figure 5-2 a and b.** Two cross sections of Japanese quail eggshells, magnified 250 times. (a) The bird that produced this egg has been fed a diet containing DDT. The structural weaknesses caused by DDT make this egg quite fragile. (b) The diet of the bird that produced this normal egg did not contain DDT. *(Wide World Photos)*

appears to be in an increased breakdown of steroids promoted by the organo-chlorine insecticides. However, there are some interesting differences and, in general, the case of birds is much more dramatic with a number of species becoming gravely endangered owing to an inability to either breed or to raise viable young. The species of birds most affected are those highest on the food chain (see Chapter 2), due to the tendency of the highly fat-soluble insecticides to become concentrated in the fats of organisms to a progressively greater extent as one advances up the chain. Thus, the species of birds especially endangered from pesticide toxicity are those feeding primarily on fish or other animal sources. These include some diving birds such as pelicans, and raptors such as falcons, ospreys, and eagles.

While there appear to be important changes in bird fertility after administration of organo-chlorine pesticides, the most dramatic alterations are connected with the famous "thin eggshell phenomenon." This phenomenon reflects a defect in calcium deposition leading to the formation of defective eggshells which leads, in turn, to drastically reduced viability and hatching.

It is significant that the deposition of calcium and calcium metabolism in general are controlled by steroid hormones and are closely coordinated with the reproductive pattern of birds. In this case, the significant steroid hormone is probably *estradiol*

which is important in the regulation of various aspects of sexual function. It has been shown that administration of small doses of DDT and related compounds leads to an increased breakdown of estradiol and a decrease of its concentration in the blood. This decline appears to result in a number of abnormalities in reproduction, including thin eggshells, late or nonexistent breeding, fewer eggs laid, aberrations in breeding behavior (including the eating of eggs), and a lower rate of embryo survival.

It is interesting that, although calcium deposition in eggshells is defective, there is apparently no block in dietary intake of calcium which is under regulation by another steroid, vitamin D.

Thus, the effect on estradiol seems to be quite specific and related to reproduction per se. For instance, measurements of calcium deposition into bone, made using the radioactive form of calcium $^{45}Ca$, show a normal picture even after administration of DDT.

It is an understatement to say that all features of the effects of insecticides on reproduction of birds have not yet been determined. The relations between hormone level and effective breeding are not completely understood. Moreover, it has been suggested that there may be other effects of the insecticides which are of equal importance and unrelated to steroid metabolism. For example, the enzyme carbonic anhydrase, which catalyzes the reaction

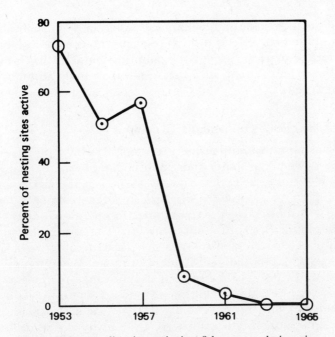

**Figure 5-3.** Decline in perigrine falcon populations in the northeastern United States, 1953–1965. (Data from W. R. Spofford in J. J. Hickey, *Peregrine Falcon Populations: Their Biology and Decline,* 1969, The University of Wisconsin Press, p. 178, table 14.1.)

$$CO_2 + H_2O \rightleftharpoons H_2CO_3$$

may play a role in deposition of the insoluble calcium salts of eggshell; the thin shells may result from a lower activity of this enzyme in the shell gland where deposition occurs. It is therefore interesting that this enzyme exhibits lower activities in both blood and shell gland in birds to whom organochlorine insecticides have been administered.

Apart from effects of insecticides on breeding and fertility, birds respond to much larger concentrations in a more dramatic way, namely by dying outright. It is in the nature of insecticide use that application is often highly localized. When elms are sprayed for Dutch Elm Disease, very high local concentrations of insecticide result, and there are many reports of massive declines in population of such insectivorous birds as the American robin. These declines appear to reflect outright killing of birds, probably due to the eating of contaminated insects and subsequent interference with the normal operation of the nervous system. Such effects require enormous doses and often a rather careless approach to application. The more subtle effects of the insecticide on breeding success occur at much lower levels. In addition, these toxic effects of high concentrations of pesticides are quite specific for the pesticide used. For example, when DDT has been replaced with methoxychlor for elm spraying, instances of high mortality have been greatly reduced.

## Effects of Insecticides on Aquatic Organisms

Some of the most profound insults to ecological stability produced by insecticides occur when the compounds enter bodies of water. For instance, the population of landlocked salmon (*Salmo sebago*) in certain lakes in the northeastern United States declined disastrously a few years ago. This decline coincided with the discovery of substantial concentrations of DDT and other insecticides in the flesh of the surviving fish. The source of contamination was easily traced to the extensive spraying of the surrounding land from airplanes as part of a mosquito abatement program. When the program was itself abated, the insecticide levels fell and the salmon population made an impressive recovery. Many similar cases have occurred, and it develops that fish are uncommonly sensitive to organo-chlorine insecticides.

Other aquatic organisms exhibit great sensitivity to insecticides. A number of marine and freshwater invertebrates have been shown to concentrate DDT from the environment and to experience deleterious effects from it. In the case of the American lobster (*Homarus americanus*) a high degree of sensitivity is not surprising, owing to the close evolutionary relationship between the lobster and the insects that the chemicals are supposed to kill. In this instance, it is probable that the action of DDT on lobsters is like a nerve poison just as it is in the case of insects. In other instances, the precise

Table 5-4. Effect of insecticides on fish.

| Species | Concentration required to kill 50% in 90 hours (expressed in micrograms per liter) | | |
|---|---|---|---|
| | Aldrin | DDT | Chlordane |
| Sunfish | 13 | 16 | 22 |
| Goldfish | 28 | 27 | 82 |
| Rainbow trout | 18 | 42 | 44 |

Source: R. D. O'Brien, *Insecticides—Action and Metabolism* (New York: Academic Press, 1967).

action of insecticides on aquatic organisms is less obvious. A number of marine algae are sensitive to insecticides to the extent that addition of DDT to mixed populations of algae leads to a dramatic shift in the proportions of the different species present. The nature of the suppression of the sensitive species is not yet understood, but it is nonetheless obvious that great ecological dislocations are possible.

The great sensitivity of fish to organo-chlorine insecticides appears to originate in the membrane physiology of the animals. Fish, whether in freshwater or saltwater, regulate their internal salt concentration by pumping ions across the membranes that form an interface between their cells and the surrounding water, notably those of the gills and the intestinal tract. We noted in Chapter 3 that a great amount of such "pumping," or active transport, occurs by means of the sodium-potassium ATPase which couples the movement of sodium and potassium to the hydrolysis (breakdown) of adenosine triphosphate (ATP), the so-called high energy compound.

There is now much evidence that DDT and related compounds inhibit the sodium-potassium ATPase of fish even when present at very low concentrations. This inhibition has been demonstrated with ATPase preparations from intestinal lining and gills from a wide selection of marine and freshwater fish. Since it has also been shown that equally low concentrations of DDT disrupt the osmotic balance of eel intestine—that DDT leads to an altered balance of salts and water—it seems that the high toxicity of DDT in fish is a consequence of the inhibition of the ATPase. One should remember that the ATPase is a transport protein located inside (or on) the lipid-rich cellular membrane where DDT and similar compounds would be expected to become concentrated. Thus, events in and around membranes would be expected to be subjected to higher concentrations of DDT than those elsewhere in the organism.

Finally, one should recall that the primary action of DDT on insects is at the level of cellular membranes, namely those of the nerve cells, as it is in the case of fish. In both instances, there seems to be an important inhibition of some function connected with membrane transport, and one might ask whether there is any similarity in the mechanisms involved. Unfortunately, our present level of understanding of the inhibitory actions involved is insufficient to form any conclusions to this interesting question.

# Chapter 6

# Atmospheric Pollution

The hand of industrial man upon this earth is nowhere more evident than in the mantle of haze that frequently obscures the sky across much of its surface. Considering that we and other land animals breathe huge volumes of air to obtain oxygen for tissue respiration, it is not odd that air pollution is of concern to us. What is perhaps more surprising is that our species continues to thrive (or to appear to thrive) in the great industrial cities where foreign material in the atmosphere reaches catastrophic levels. In fact, with a few exceptions, cases in which atmospheric pollution has been shown to contribute to cellular injury remain rare, and levels of contamination that are quite unacceptable from the esthetic viewpoint appear nonetheless to be relatively harmless. An interesting case in point is a study that was performed in London several years ago, where the health of workers in bus garages was monitored. It developed that the health of workers in diesel-engine garages where carbon monoxide and hydrocarbons were at high concentrations in the air was nevertheless superior to that of workers in garages where all vehicles were electrical. This difference was the result of a policy that allowed smoking in the latter but not the former locations.

There is little mystery about how the atmosphere becomes polluted. Large cities bristle with smokestacks and even the best automobiles produce large quantities of particulate and gaseous waste. In the last instance, the impact of automobiles is multiplied by their large numbers, their inherently wasteful use patterns, and the widespread practice of doping their fuel with compounds of unoxidizable lead.

One may consider air pollution with respect to the compounds involved and speak of two broad classes of materials emitted. First, there are materials delivered to the atmosphere which are normal in the sense that they would be present without the agency of man. Such compounds could not be called inherently foreign or even harmful, except as their concentration exceeds normal limits. A major product of most forms of combustion, including that of gasoline and other fossil fuels, is carbon dioxide ($CO_2$) which is also a product of normal animal respiration (itself a form of combustion, in a

**Figure 6-1.** Dense smog over Los Angeles Civic Center.

larger sense). Not only is $CO_2$ a normal constituent of the atmosphere, but it is essential as a primary carbon source for photosynthetic plants. It even plays a role in the nutrition of nonphotosynthetic organisms, including humans, through a number of reactions leading to the "fixation" of $CO_2$ to produce metabolically-useful compounds. Similarly, carbon monoxide is a normal constituent of the atmosphere, formed by radiation-induced reactions in the upper atmosphere as well as some processes in the biosphere, such as the metabolism of certain algae. Thus, the presence of neither compound could be termed abnormal per se, and difficulties from our point of view arise only when concentrations exceed certain limits. These limits are very different for the two compounds in question, with the monoxide compound being, of course, of much greater toxicity. Since more carbon monoxide is emitted by inefficient combustion processes (it represents only a partial oxidation of carbon compounds), there is a premium on improving the efficiency of engines, power generation, etc., so that the product will be the much less toxic $CO_2$. This is a felicitous instance where economic self-interest, or efficient operation, and environmental safety run in parallel, an occurrence not nearly as rare as some people like to imagine.

In contrast to carbon dioxide and carbon monoxide, many compounds produced by industry would not be found in the atmosphere in measurable quantities unless humans placed them there. Many hydrocarbons are in this category as are products of many specific industrial processes and, of course, the dreaded tetraethyl lead. In these instances, organisms are confronted with a gaseous phase containing compounds that are utterly foreign to the biosphere, and their toxicity often reflects their foreign nature.

In reflecting on the harmful effects of pollutants, especially those in the second, or foreign, category, one notes that there are great differences in the subtlety of their action. For instance, we shall consider the influences of carbon monoxide on respiratory processes, both in the context of transfer of oxygen by blood and, on the cellular level, in the inhibition of cell respiration. These effects are very specific in that the compound interacts with a few highly specific cellular constituents. This specificity and delicacy are in striking contrast to the action of a number of other compounds also found in polluted atmosphere. Thus, some other forms of pollution are harmful for reasons not unlike hitting a cell with a hammer. Particulate material often simply clogs respiratory passages in animals. This is not a very subtle effect, but sufferers of some of the great London smog episodes of the last decade would fail to appreciate that distinction. Likewise, many compounds that emanate from smokestacks are harmful to plants and animals simply because they are highly corrosive. Cell biology is only minimally involved in understanding why sulfuric acid in the air is harmful, and the effects of many materials do not progress beyond their corrosive influence on the linings of lungs and air passages or the surface layers of plants. To state it differently, when air pollution removes the paint from houses and corrodes the stone of ancient churches, its harmful effect on creatures is anything but surprising and largely fails to excite our intellectual curiosity.

## Physiological Effects of Carbon Monoxide

We have seen that carbon monoxide is a product of incomplete combustion of carbon compounds (and all combustible fuel is in this category) and that its toxicity has an effect on cellular respiration. A truism appears frequently in this book to the effect that compounds which mimic those in organisms tend to be toxic by competing with the normal ones in whatever processes they participate. Carbon monoxide is an excellent example of such molecular imitation, and the compound that is imitated in this case is molecular oxygen. The two compounds exhibit a rather similar structure, with oxygen existing as the diatomic molecule, $O_2$, which is similar to the diatomic carbon monoxide molecule, CO.

In this instance, the point of competition between the two compounds is their binding to certain iron-heme-protein compounds, namely, hemoglobin in the red blood cells (erythrocytes) and cytochrome oxidase in the mitochondria of aerobic cells from higher organisms as well as the plasma membranes of many bacteria. First we will consider the effect of carbon monoxide on the hemoglobin of blood.

Hemoglobin is a protein that functions to transport oxygen from the lungs or gills to the tissues of the body, where it is consumed through cellular respiration. Hemoglobin consists of a protein portion to which are bound four iron-heme groups with the following generalized structure:

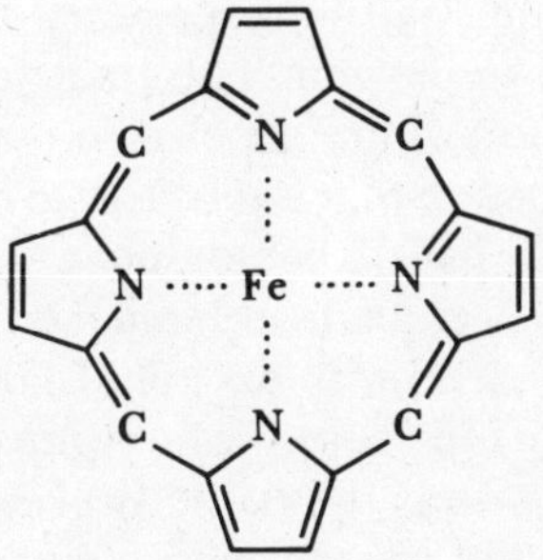

More precisely, hemoglobin consists of four protein subunits (of two somewhat different types) with each bearing its own heme unit. The functional form of hemoglobin is thus the complete four-protein, four-heme compound, and the whole structure is able to bind four molecules of oxygen simultaneously. The molecule absorbs oxygen at the lung or gill capillaries where the oxygen concentration is high, and it deposits oxygen in the capillary bed of other tissue where its concentration is low. It is thus an oxygen transporter, and interference with its orderly function leads to disastrous effects.

Carbon monoxide interferes effectively with the oxygen translocation function of hemoglobin. It reacts with the iron-heme system in the same manner as oxygen and therefore competes with oxygen for a place on the carrier. In fact, it is more correct to state that the oxygen (or carbon monoxide) forms a coordination compound with the reduced iron of the iron-heme system and that when an iron is occupied with a carbon monoxide molecule, it is unable to accommodate an oxygen molecule. To say that carbon monoxide competes with oxygen for a site on the hemoglobin molecule rather understates the situation, since their relative affinities (tightness of binding) are not at all identical. Thus, the affinity of hemoglobin for carbon monoxide is about two hundred times as great as that for oxygen. This greatly favors carbon monoxide in the competition and insures that rather low concentrations of the monoxide can be quite toxic.

A final feature of the carbon monoxide-hemoglobin interaction is that the complex is light-sensitive. In other words, the complex may be broken down by a photochemical reaction. A similar situation pertains with other carbon monoxide-heme compounds and will be discussed below.

It is clear that low concentrations of carbon monoxide are immensely harmful in that they displace oxygen on blood hemoglobin molecules, thus starving the tissues for oxygen. In fact, the carbon monoxide is transported to the tissues in place of oxygen and it is here that a second interference with respiration occurs. As we saw in Chapter 3, animal cells require oxygen in order to oxidize foodstuffs, a process yielding energy in the form of ATP. Respiration amounts to removing electrons from molecules derived from food and transferring them via the respiratory chain (flavoproteins, cytochromes, etc.) and finally to oxygen. The final reaction of the chain is the cytochrome oxidase reaction where the electrons are used to reduce oxygen to form water. It is this reaction that is sensitive to carbon monoxide, and the sensitivity is analogous to that of hemoglobin. Thus, cytochrome oxidase (also called cytochrome $a$) is an iron-heme-containing protein, and the monoxide competes at the iron with oxygen. In other words, not only does carbon monoxide choke off the supply of oxygen to the tissues, but (adding insult to injury) it then poisons the tissue process that requires oxygen to begin with. The net effect of carbon monoxide is to prevent the synthesis of ATP that is coupled to electron flow in the cytochrome system. When the energy source (ATP) is no longer synthesized, cells are able to survive only a matter of seconds. Because the brain requires a particularly active ATP synthesis, it is one of the first tissues to be damaged and, of course, its damage immediately produces major and usually irreversible damage to the rest of the organism, such as cessation of breathing and circulation.

It is thus no mystery why enough carbon monoxide will kill a human. However, some interesting questions remain to be answered concerning the effects of less than lethal doses of the gas. It is well known that smaller amounts of carbon monoxide can produce headaches, mild confusion, and some respiratory symptoms. What is not clear is whether these levels of toxicity will, over a long period of time, produce major damage. This is an important point since many people work under conditions where they breathe air with considerable amounts of the gas. Such conditions are especially severe where gasoline engines are run with inadequate ventilation, such as in vehicle tunnels, at turnpike toll booths, and at bus depots.

## The Influence of Carbon Monoxide on Plants

Since oxygen transport and cell respiration are the sites of carbon monoxide toxicity, it might be expected that organisms which either do not carry on these processes or do but are not totally dependent on them for energy, would be insensitive to the poison. In fact, this is generally true. Green plants which are photosynthetic, while carrying on respiration as well, are not as dependent on oxygen and are generally less sensitive to carbon monoxide. Of course, some tissues of plants such as roots do not obtain energy from light because they are underground, and these are more sensitive to respiratory inhibitors. Some marine algae exhibit a rather extreme insensitivity to carbon monoxide: they synthesize the gas and use it to fill flotation bladders. One wonders, in this case, whether the role of the monoxide is that of a gaseous antibiotic; the bladders would certainly be an inhospitable environment for any aerobic pathogenic bacterium finding its way there. It is also puzzling how such algae function in the dark. Most plants contain mitochondria which, in turn, contain cytochrome oxidase, and these would be expected to be sensitive to carbon monoxide.

A second process carried out by certain groups of plants is of worldwide ecological significance and has been shown to be sensitive to carbon monoxide. Many plants, especially those of the pea family, are able to incorporate nitrogen from the atmosphere into their constituent compounds. This process, called "nitrogen fixation," is an important part of the nitrogen cycle. Plants that carry it out do so in combination with certain bacteria residing in root nodules, most of which belong to the genus *Rhizobium*. This cooperative arrangement is a true symbiosis in that neither the bacterium nor the plant alone can carry out the whole process. Interestingly, the specialized nodules where the bacteria occur contain hemoglobin, a molecule otherwise confined to the animal kingdom. In this case, the role of the hemoglobin seems to be involved with the regulation of the free oxygen concentration in the nodules by binding a variable amount of it. Without considering the question of how nitrogen fixation works, it is of interest that it is competitively inhibited by carbon monoxide. The details of this inhibition are not completely determined, and the mechanism of nitrogen fixation itself is still obscure so that one cannot assess the ecological significance of the inhibitory process. At our present level of knowledge, it can only be noted that nitrogen fixation is yet another point in the ecosystem where the gaseous pollutant, CO, can in principle have a deleterious effect. In this last association, it is also of interest that the inhibition of nitrogen fixation by CO reflects a more general aspect of inhibition, namely, that a variety of gases compete with nitrogen. These include hydrogen ($H_2$), nitrous oxide ($N_2O$), and nitric oxide (NO), the last two being components of industrial atmospheric pollution.

## Effects of Some Nitrogen Compounds

We have just seen that certain oxides of nitrogen (NO and $N_2O$) are inhibitors of nitrogen fixation. In fact, these compounds, which are important

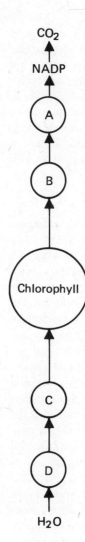

**Figure 6-2.** A schematic view of photosynthesis. The arrows denote electron flow analogous to that of respiration (see Figure 3-8). Chlorophyll, when stimulated by absorbed light, serves as the electron pump. The source of the electrons is the water molecule, which is converted to free oxygen plus hydrogen ions upon removal of the electron. A through D denote the carriers of the photosynthetic electron transfer chain, which are actually more numerous than shown here and which include cytochromes. Finally, electrons reduce NADP to NADPH which, in turn, serves to reduce carbon dioxide ($CO_2$) to cellular components such as sugars and other compounds.

components of pollution, produce a number of effects on cells of both plants and animals. Unfortunately, the mechanisms underlying these effects are often obscure so that we can generally only list them with little comment. One practical difficulty in studying the influences of these gases is the regrettable ease with which nitric oxide (NO) is converted to $NO_2$ (nitrogen dioxide), so that one is not always sure which compound is being studied.

Small levels of oxides of nitrogen in the atmosphere injure sensitive plants, producing, for instance, visible injury to leaves in a short time. The underlying cellular basis for the injury may be complex, but it is of significance that these gases dramatically inhibit photosynthesis long before visible damage occurs. Thus, a reasonable possibility would be that these gases may be considered direct inhibitors of the photosynthetic process and that the inhibition may account for much of the generalized damage. It is not certain at what point in the photosynthetic process these compounds act; indeed, there may be more than one locus. One intriguing possibility is related to the role of nitrate ($NO_3-$) in normal photosynthesis. It is recalled from Chapter 3 that photosynthesis occurs when electrons are "pumped" by chlorophyll which has been excited by light. In most cases in higher plants the electron comes ultimately from a water molecule and travels through the chlorophyll "pump" along a series of cytochromes and other electron carriers, and it ends up by reducing a final electron acceptor (see Figure 6-2). Most commonly the final acceptor is $CO_2$ and its reduction leads, by a complicated metabolic route, to sugar molecules and other cell components. Indeed, the "fixation" of $CO_2$ to form various cellular molecules was once considered to be the essence of photosynthesis, although we now know that it is one of several results of the photosynthetic process. In addition to $CO_2$ serving as final electron acceptor, plants also can reduce nitrate ($NO_3-$); this process is of great importance in providing nitrogen compounds that are cellular constituents. The photosynthetic reduction of nitrate is also an important link in the nitrogen cycle.

The point of the preceding discussion is the probability that the part of photosynthesis most sensitive to nitric oxide and nitrous oxide is the location at which the final electron acceptors become reduced. Some evidence exists for this idea and examination of the structures of the molecules involved support it, so that competition *could* occur between the nitrate ion ($NO_3-$) and the rather similar nitrous oxide gas ($N_2O$). In fact, it is probable that nitric oxide (NO) is a competing molecule as well, and both may also compete with carbon dioxide. Remember in all cases that when we speak of competitive inhibition, we are really referring to competition for a site on some enzyme or other binding protein. Thus, we interpret the action of nitrous oxide on photosynthesis as sufficiently resembling either nitrate or carbon dioxide to simply push them aside and replace them at the active site, where their electron acceptor function occurs.

The effects of oxides of nitrogen on animal cells are similar to those on plants only in that they are complex. Obviously, the exact point of attack on animal cells must be different since animals do not carry out photosynthesis. Before discussing the nature of physiological harm to animal cells, it is worth documenting the fact that such harm does occur. In the first place, nitrous oxide (as opposed to nitric oxide) appears to be more harmful to animals and, in particular, to humans. Indeed, the chief physiological contribution of nitric oxide is to be a ready source of the other compound. Nitrous oxide (also formerly known as laughing gas) can produce profound physiological effects on humans as the compound was used for many years as a general anesthetic, causing one to lose consciousness at a high concentration and to become insensitive to pain at a somewhat lower level. Such a property implies a rather dramatic physiological influence, although it must be said that anesthesia (and induction of) occurs at concentrations *much* higher than those reached in even the most vilely polluted air. In other words, anesthesia involves a high dose over a short time period; we are more interested in the chronic (long-term) effects of relatively small doses.

When animals are exposed to nitrogen oxides over any period of time extending beyond a few hours, the primary site of damage seems to be the lungs which, naturally, comprise that tissue most directly confronted by the gases. Damage includes inflammation and increased susceptibility to infection. When high concentrations of nitrous oxide are employed, for example, in the range of 0.01%, death occurs through a total collapse of lung function. Indeed, since $NO_2$ produces a situation indistinguishable from emphysema in rodents, it has been implicated in the origin of this disease in humans as well. Effects of the gas are also noted in other tissues such as liver and red blood cells. In all cases, the primary underlying cause of harm seems to be the ability of the gas to act as an oxidizing agent and, more specifically, to oxidize some of the fatty acids of membranes (see Chapter 3). The oxidizing ability of nitrous oxide is also able to oxidize hemoglobin to a nonfunctioning derivative called methemoglobin, in which the central iron is oxidized (from the ferrous to the ferric state). In addition, nitrogen oxides are able to bind to normal hemoglobin (in place of oxygen) to form, for instance, NO-hemoglobin. However, it seems that this reaction is of little practical importance at the concentrations of the gas actually found in polluted atmosphere.

## Effects of Other Oxidizing Agents

A number of other oxidizing agents are formed by the action of sunlight on components of polluted air. Because such photochemical reactions are extremely complex and, in some cases, not well understood, we shall give only an example in brief outline. Nitrogen dioxide ($NO_2$) in the air can be cleaved in the presence of light by the following reaction:

$$NO_2 \rightarrow NO + O.$$

The NO (nitric oxide) that is formed is rather harmless and has been discussed previously. The oxygen, however, is a different case. Normal oxygen in the atmosphere is in the form of a molecule, $O_2$, consisting of two atoms. Single atoms of oxygen are rare and highly reactive. They can, for instance, combine with organic molecules to form organic peroxides, an example of which is methyl-peroxide, $CH_3OOH$, a very reactive and strong oxidizing agent. Another set of reactions involving an organic molecule and another $NO_2$ molecule can yield peroxyacyl nitrates of the general form, $ROONO_2$, where the R is the organic part, such as methyl, $CH_3$. These are also strong oxidizing agents and are important components of atmospheric pollution, especially smog. Finally, the atomic oxygen can combine with a normal oxygen molecule ($O_2$) to form $O_3$, which is ozone.

Most of the remarks just made about the effects of nitrogen oxides also apply to these other oxidants. The primary site of action is, predictably enough, the lung, and the cellular membranes appear to be most drastically affected with oxidation of fatty acids and certain membrane proteins, too. Because the sensitivity of lungs to oxidants has especially serious consequences for people with other lung diseases, serious smog episodes have produced many fatalities in cases of asthma and emphysema.

## Some Organic Components of Polluted Air

A great amount of organic material (compounds containing carbon and hydrogen) enters the atmosphere as a result of incomplete burning of various fuels. Since these fuels frequently confer an offensive odor on polluted air, they represent an important part of the esthetic problem of air pollution. On the other hand, any physiological harm resulting from inhalation of these materials is so subtle as to be quite immeasurable in all but a few cases. For instance, hydrocarbons (compounds containing *only* carbon and hydrogen) are remarkably biologically inert and have not been shown to be harmful to animals except at levels much higher than those encountered in even the most severe pollution. It is true, however, that some specific organic compounds do produce specific harm. For example, organic lead compounds exhibit the characteristic toxicity of lead and may be considered as vehicles for transmitting biologically-soluble lead to air-breathing animals. (These compounds are discussed in the chapter on heavy metal pollution.) Certain other organic compounds are toxic for one reason or another. There are, for example, a number of aldehydes, compounds with the function group —CHO, which are irritants even at the low concentrations found in air. However, these are rather exceptional, and hydrocarbons and many other organic compounds are, as we said, remarkably inoffensive.

Of course, we have seen one case where even rather inoffensive compounds produce difficulties by serving as starting materials for subsequent reactions. Thus, the photochemical production of organic peroxides can begin with organic compounds which are themselves physiologically inert. It is also sometimes true that certain compounds can produce specific effects on some organisms but not on others, owing to some particular aspect of the physiology of those organisms. For instance, a common component of auto exhaust and other exhausts is the gas ethylene, $CH_2$—$CH_2$, which is harmless to animals unless they practically bathe in it. On the other hand, this compound is highly toxic to a variety of higher plants. It appears that the reason for harm is the role of ethylene as a normal growth-regulating substance in these organisms. In this case, the harm produced by ethylene occurs through an overriding of normal controls on growth so that unregulated growth occurs, which is disastrous to the plant. This is reminiscent of the use of plant-growth hormone analogs such as 2, 4-D as herbicides, their basis of action being that unregulated growth is quickly fatal.

# Chapter 7

# Metal Pollution*

Chemical analysis of plant and animal tissue reveals two categories according to which all of the elements may be classified. In the first category are the major constituents including hydrogen, carbon, oxygen, nitrogen, phosphorus, calcium, potassium, magnesium, chlorine, and sulfur. The total contribution of these elements represents 97-98% of the organism's weight. The remainder of the elements are known as *trace elements* because in the early days of biological chemistry it was not possible to express their exact concentration owing to technical limitations, so their concentrations were listed as a "trace." Since then accurate techniques have been developed to measure these elements, but the term persists.

Trace elements may be divided into three categories (see Table 7-1). The first group includes those known to be essential for normal plant and animal function since their removal from the nutrient source causes structural and physiological abnormality. For example, a lack of iron is manifested in humans by defective red blood cells which are fewer in number, smaller in size, and contain a lower concentration of iron. These are symptoms of iron-deficiency anemia about which television commercials regularly remind us.

The second group includes several elements which are considered essential but about which absolute evidence is still insufficient. Fluorine is a good example of this situation. This element appears to aid in prevention of dental problems and in promotion of development of normal skeletal structure, but a diet deficient in fluorine has not been unambiguously shown to limit growth or produce other abnormalities.

The third group includes elements commonly found in plants and animals but which have not yet been shown to be essential or beneficial in any way. These are thought to be present simply because they appear in the chemical environment of the organism. This group exhibits a wide range of concentration in plants and animals. This situation contrasts with the essential trace elements which are present in much narrower concentration ranges and are less dependent on variations in their nutrient source.

---

*By C. T. Settlemire, Departments of Biology and Chemistry, Bowdoin College.

Table 7-1. The classification of trace elements.

| Group 1 essential | Group 2 possibly essential | Group 3 nonessential |
|---|---|---|
| iron | nickel | aluminum |
| iodine | fluorine | antimony |
| copper | bromine | mercury |
| zinc | arsenic | germanium |
| manganese | vanadium | silicon |
| cobalt | cadmium | rubidium |
| molybdenum | barium | silver |
| selenium | strontium | gold |
| chromium | | lead |
| tin | | bismuth |
| | | titanium |

Many of the elements in this third group are well known because of their toxicity, as in the case of lead and mercury. However, all of the trace elements, including the essential ones, are toxic if ingested for a long enough period or at sufficiently high concentrations. It is important to remember that if "a little bit is good, more will not necessarily help," and in fact may be quite toxic.

The primary purpose of the remainder of this discussion is to consider examples of the important nonessential elements which we regard as contaminants in our human environment and to examine their physiological effects. We will examine the origins of such pollution as well as the manner in which they may interact with plant or animal systems.

## Mercury Poisoning

Mercury poisoning has become a matter of public awareness as the result of occurrences in recent years. One of the first major cases to be publicly recognized occurred in Japan's Minamata Bay region where an unknown illness appeared in 1953 and caused many fatalities. People did not associate it with mercury poisoning until ten years later, and eventually 111 cases were identified. Since then, concern has grown in Sweden, Canada, the United States, and elsewhere, primarily directed toward the distribution of mercury in water sources and its probable effects.

Mercury is one of the less abundant elements on the earth's surface. It comprises less than $3 \times 10^{-6}$ percent of the crust and is generally found in combination with other elements. It has been mined on all continents except Antarctica, and some of the earliest mining operations date nearly 3100 years ago in China. In almost all cases, the ore of interest has been cinnabar, a combination of mercury and sulfides.

The metal exhibits several properties which were of interest initially because of their unusual nature and, more recently, because of their fulfilling several industrial purposes. These properties include the following:

1. Mercury is the only metal that is a liquid at room temperature, not freezing until a temperature of $-39°$ C. is reached. This property is important in cases where mercury serves as a chemical catalyst in industrial processes.
2. Many other metals dissolve in mercury to form useful alloys.
3. Mercury is an excellent conductor of electricity, allowing it to be used in electrical switches as well as in lamps and batteries. Its electrical properties have also led to its use in the production of important industrial chemicals such as sodium hydroxide and chlorine gas. Both are produced by electrolytic processes—processes in which electrical current flows through a chemical system causing the cleavage of the chemical into positive and negative ions which then migrate toward the appropriate electrodes. In such applications, metallic mercury is used to form one of the electrodes and, in fact, more mercury is used for this purpose than any other.
4. Mercury and compounds containing it are well known to be very toxic. Such toxicity has led to its wide use as a pesticide. In agriculture, compounds of mercury are often employed as fungicides, either as coatings on seeds or on growing plants. The fungicidal properties of mercury compounds are also the basis for its use in the paper industry where it can prevent the growth of moulds on wet paper materials. Mercurials have also been added to paints as a fungicide to prevent mildew and as an anti-fouling agent in boat paints to prevent growth of barnacles and other marine organisms.
5. Mercury is an effective catalyst in the manufacture of several commercially important products including the manufacture of vinyl chloride from acetylene. Vinyl chloride is a central material in the production of plastics. It was, in fact, a factory producing vinyl chloride that was dumping mercury into Japan's Minamata Bay.

Man's many uses of mercury have resulted in a redistribution of this element with much higher concentrations in his immediate environment. Mining operations have exposed unclaimed mercury to the action of wind and water and have resulted in leaching into rivers, lakes, and the oceans, and vaporization into the atmosphere. Runoff from agricultural uses and the dumping from factory operations have been major contributors to the addition of mercury to air and water. Practically no attention was given to levels of mercury in the aqueous environment in the United States before 1970. As a result, it is impossible to determine what the "natural" mercury content of water might have been. One might expect mercury, at least theoretically, to be distributed throughout our atmosphere and water supplies since mercury has a high vapor pressure, that is, it vaporizes readily and can be expected to diffuse rapidly from a site of introduction.

As we discussed earlier, nonessential trace elements tend to accumulate in organisms according to the concentration in their nutrient and water source. Mercury follows this pattern, as demonstrated by laboratory feeding experiments. One of the important questions posed in the last few years has been to what extent levels of mercury have increased in living organisms. While comparisons with analysis of plant and animal materials in the 1930s are of some value, analytical techniques have improved so that one might question the accuracy of the older analysis. Suffice it to say that during recent years there have been measurable increases in mercury content of a number of organisms, although it is not certain what constitutes a dangerous level.

Mercury (Hg) is used in the form of its metal, as inorganic salts, and as organic compounds. Examples of inorganic salts include mercuric chloride ($HgCl_2$) which is an important catalyst in plastic production and mercuric oxide (HgO), an anti-fouling agent. Organic mercury compounds are used primarily as fungicides. Some examples include phenyl mercuric acetate (PMA), a major fungicide used by the pulp and paper industry.

$$\langle \bigcirc \rangle HgO-\overset{\overset{O}{\|}}{C}-CH_3$$

Other common organomercurial compounds also have their major use as a fungicide in the treatment of seed grains. They have either a methyl ($CH_3-$) or an ethyl ($CH_3CH_2-$) group attached to mercury plus some other organic portion. Examples include methyl-mercury-acetate,

$$CH_3-Hg-C\underset{OCH_3}{\overset{O}{\diagup}}$$

methyl-mercury-nitrile,

$$CH_3-Hg-C\equiv N$$

and ethyl-mercury-chloride.

$$CH_3CH_2-Hg-Cl$$

The effect of including an organic portion in these mercury compounds is an increase in the solubility of the attached mercury in the tissues of organisms. Because the compounds are much more soluble in the fats of cellular membranes, they are much more able to enter cells and exert a toxic effect. The increased toxicity of the organic compounds for fungi is obviously the basis for their use as fungicides, and it is evident that they are much more toxic to other organisms for the same reason. Thus, levels of mercury in the environment which are unexceptional when the mercury is inorganic become causes for extreme concern when the major forms are the organic derivatives.

When high levels of inorganic mercury are present in the environment, however, increasing amounts of it are converted by microorganisms into organic derivatives such as dimethylmercury

$$(CH_3)_2 \quad Hg$$

which are much more toxic to higher organisms. It is possible that the reactions in bacteria leading to this methylation are detoxification reactions benefiting the bacteria. But if that is true, their detoxification is some other organism's anathema since it appears that much of the organic mercury which is found in fish from mercury-polluted waters comes by this sort of process.

Considerable information is available on possible mechanisms for the harmful effects of mercury on cells. It is well known that mercury (as well as other heavy metals) binds to the sulfhydryl (—SH) groups of proteins and that binding can alter the catalytic properties of enzymes. A basis for action is thus available, although there is considerable question as to what proteins and what functions are targets at low concentrations of the poison. At the present time, the most likely candidate for primary action of mercury must be the membranes of cells, especially the external plasma membrane. Numerous studies have shown that mercury binds to these membranes in various cell types and that it interferes with a number of membrane functions including transport of various ions and molecules. For example, mercury renders cellular membranes much more permeable to potassium and hydrogen ion and much less permeable to various nutritionally important molecules such as sugars and amino acids.

Many of the human tissues most sensitive to mercury are those in which membrane transport is of special functional significance. For example, kidney function is impaired with increased water loss, and the dysfunction is related to both the central role of membranes in the kidney filtration process and to the likelihood that mercury will be deposited there as the organism endeavors to secrete it. The nervous system is also a site of mercury toxicity; the reader will recall from Chapter 3 that ion movement across membranes provides the basis for nerve transmission.

The major difficulty in being more specific about the target of mercury in producing toxic effects is that virtually all cellular functions either involve enzymes with —SH groups or membrane surfaces that, likewise, bristle with the same groups. Thus, there are simply too many possible candidates, and the details of mercury interaction remain far from being ascertained. Although the plasma membrane is, for obvious reasons, more exposed to mercury (and all other environmental materials), there are other membrane systems within the cell that may also be sensitive. For example, mercury compounds have been shown to inhibit ATP synthesis occurring in isolated animal mitochondria.

## Lead Poisoning

Lead has been an important metal in industry since early recorded history, and its toxic properties have been known for almost as long a period. It is currently used in the metallic form for many construction functions, in the production of batteries, as a base in the paint industry, and as a number of organic compounds, most commonly as a gasoline additive.

Lead enters the biological portion of the world by several routes. One route is simple industrial dumping of waste and another is the deterioration of lead paints which can free the lead to be incorporated by some organism. A route that is of public health concern is the ingestion of lead by small children in the form of flakes of lead paint. This sort of behavior—i.e., eating nonfood materials—is called *pica* and is especially pronounced in undernourished children, thus adding the risk of chronic lead poisoning to the reality of poverty and hunger. Lead has also traditionally been a component of pottery glazes, and high-lead glazes have given rise to toxicity in

Table 7-2. Origin of lead in the atmosphere.

| Source | Percent |
| --- | --- |
| coal combustion | 0.5 |
| lead-based industry | 1.3 |
| gasoline combustion | 98.2 |

people who habitually drink beverages (especially acidic ones) that have been kept in such containers.

In the United States, hunters annually deposit tons of lead in the form of shot and bullets into the environment. Shot, in particular, leads to entry of lead into the biosphere since waterfowl that survive being hit often eat considerable amounts of it as they feed on the bottom of shallow bodies of water. Since shot is often made of an alloy that contains in addition to lead a small percentage of arsenic, it can be quite toxic so that what the hunter fails to do directly, he manages by indirection.

Organic lead enters the biological world through the combustion of gasoline containing tetraethyl lead, the famous (and presently infamous) antiknock compound. Most of the lead that enters the atmosphere by this mechanism is not the highly toxic organic lead compound but, rather, inorganic derivatives such as oxides that are produced by combustion in the automobile engine. The organic forms of lead are, of course, especially dangerous because of their increased solubility in biological membrane fats in a manner analogous to that mentioned in the case of mercury compounds. It

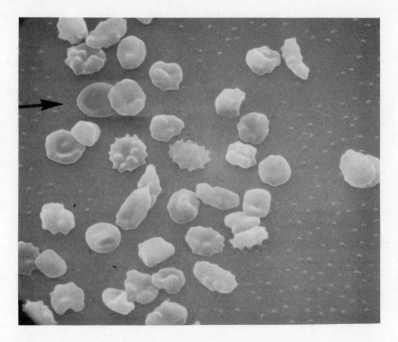

**Figure 7-1.** Red blood cells from a lead-treated mouse. The arrow points to a relatively normal cell for comparison. The bumpy cells are characteristic of lead damage. This is a scanning micrograph at a magnification of × 2000.

seems likely that the actual toxicity is exerted by lead in the inorganic form and that organic lead serves primarily as a vehicle for transporting it to its site of action.

Lead, like mercury, owes its toxicity to interaction with sulfhydryl groups (—SH) of proteins. As in the case of mercury, lead attacks functions associated with cellular membranes so that kidney and nervous systems are major sites of damage. Mitochondrial energy conservation (i.e., ATP synthesis) is impaired on addition of lead salts. Other membrane functions are also damaged, for instance, absorption of sugar and phosphate by kidney cells is inhibited as is the energy-linked transport of sodium and potassium in red blood cells by means of the $NA^+$—$K^+$ ATPase system (see Chapter 3). There is considerable evidence that the disturbance of kidney and possibly brain function by lead may reflect that element's influence on mitochondrial ATP synthesis which provides the basis for nearly all energy-requiring reactions and processes in cells.

Finally, lead poisoning leads to anemia in animals which means that there is a deficiency in the red pigment of blood, hemoglobin. This deficiency may result from three different cellular effects. First, lead toxicity leads to the production of red blood cells that are excessively fragile which, in turn, is reflected in a decreased life-span of the cells in circulating blood. Lead also interferes with the actual synthesis of hemoglobin, inhibiting an early reaction in the pathway leading to hemoglobin formation. This reaction, which is catalyzed by an enzyme bearing the same name, appears to require a sulfhydryl group for activity. In cases of animals intoxicated with lead the activity of the enzyme is low, and not only hemoglobin but other heme proteins such as the cytochromes are deficient in amount. Finally, hemoglobin synthesis is also inhibited at another locus: the incorporation of iron into the organic heme ring (see Chapter 5) is impaired and free iron tends to build up in the cells.

## Toxicity of Cadmium and Zinc

Cadmium and zinc form, with mercury, a group in the periodic table of elements possessing many similar properties because of their similar atomic structure. The close relationship between cadmium and zinc is seen in the biological world where one element sometimes protects organisms against unfavorable effects due to the other. Both are produced as the result of various industrial processes as they are involved in various aspects of the metal industry, electroplating, paint pigments, and plastics. Both are used as catalysts in chemical production with cadmium, for instance, being employed in the potentially dangerous form of diethyl cadmium.

Many aspects of the cellular toxicity of both zinc and cadmium remain to be determined. It seems clear, however, that both elements can serve as inhibitors of processes requiring sulfhydryl groups and that they produce other effects as well. Phosphate groups have been suggested as loci for these additional effects, and actual toxicity may result from a combination of both. Both cadmium and zinc appear to interact with membranes and, for example, the cation (positively charged ion) $Zn^{2+}$, alters properties of mitochondria rendering them more permeable to potassium. Zinc also inhibits the cytochrome system of mitochondria by acting at very low concentrations to interrupt electron transport between cytochromes $b$ and $c$. Both cadmium and zinc ion also interfere with the reactions of mitochondrial ATP synthesis, effectively disturbing energy conservation in aerobic cells.

Other metallic elements have been either actually implicated or suggested as having toxic effects. For example, beryllium, which is used in many alloys, has been shown to produce poisoning known as beryllium disease in humans. It is a sometimes fatal condition centered around lung damage. The cellular basis of this disease appears obscure, and it is fortunately not threatening to people who do not work in a factory that makes beryllium alloys. Other metals are probably of concern in some instances, such as common metals like iron and copper which can be toxic and may be found in polluted water. These, however, do not constitute a threat as serious or as specific as that of mercury or lead. Likewise, toxic effects have been attributed to uncommon metals like chromium and vanadium, but it would appear that large doses are required and the nature of the toxicity is, in any case, obscure.

# Chapter 8

# Additional Forms of Pollution

By now it should be obvious that there are numerous categories of pollution which can, in principle, harm cells. In fact, almost every industry that does any work with chemicals makes its own special little contribution to our environment, whether it be ejected into the atmosphere or poured into bodies of water. It is beyond the scope of any book that is not an encyclopedia to attempt examination of the cellular effects of all types of pollution. For this reason and because little is known about cellular effects of many common polluting compounds, this book has been highly selective and has dealt with examples of pollution effects rather than an exhaustive survey. This chapter presents a brief discussion of a rather random assortment of compounds that do not fit readily into the categories of the previous sections. In the following material it will become highly apparent that much is not well understood and that there is consequently a great amount to be accomplished by future research.

## PCB's: Polychlorinated Biphenyls

During the past few years, considerable interest has been generated about a class of compounds called polychlorinated biphenyls, which we shall refer to by their popular abbreviation, PCB. These compounds are by-products of the plastic industry as well as components of paint, lubricants, and electrical insulators, and they appear to have a worldwide distribution in the environment. They enter the environment both as effluent from some industrial sites and dumps and, perhaps more importantly, by being leached from plastic objects by water action and by being released on incineration of plastic materials. Plastic objects have also reached the point of worldwide distribution (as well as limited distribution on the moon), and even remote portions of the oceans are contaminated by small chunks of plastic waste.

The ubiquitous distribution of the PCB's was first apparent when these compounds appeared as contaminants in gas-chromatographic analysis for insecticides. So prevalent were they that it is now widely held that such analysis performed more than a few years ago must be regarded as virtually useless since, in most instances, the PCB's were mistaken for insecticides, thereby greatly elevating the apparent values. In fact, PCB's demonstrate a number of similarities to chlorinated hydrocarbon insecticides: they not only act like them in gas chromatographs, but also have quite similar solubilities. For this reason, they travel in the environment in a similar manner, being concentrated in the fat portions of organisms and becoming concentrated as they travel up the food chain. Indeed, like DDT, they have been detected in human fat and milk.

Relatively little is known about the action of these compounds on cells, although the present intense interest in them argues that more information will be available soon. We shall make only two comments about their cellular influence. First, the action of the PCB's appears similar, if not identical, to that of the chlorinated organic insecticides such as DDT, with the obvious exception that the PCB's are not noted for their utility as insecticides and so must exhibit important differences as well. Recent investigations of the toxicity of the PCB's in algae are highly suggestive of a common mode of action in that the toxicity of the two classes acts in a parallel manner. Thus, algae that are insensitive to DDT are insensitive to the PCB's as well, and those that are more sensitive to one are also more sensitive to the other. Since some species of marine algae (diatoms) are extremely sensitive to the PCB's, the compounds can alter the balance between various species of these organisms. Since these plants are at the beginning of the food chain (primary producers), any alterations in species composition of the ocean could produce alterations in the higher levels of the food chain so that the final effects might be profound and unexpected.

A second aspect to the influence of PCB's on cells relates to the microsomal system for the oxidation of various compounds discussed in Chapter 4. It will be recalled that various insecticides and drugs promote the synthesis of microsomal enzymes which together are able to oxidize many insecticides and drugs, as well as steroids and other compounds. It turns out that the PCB's must be added to the list of materials that stimulate formation of the microsomal oxidase systems. Such stimulation has been observed in steroid oxidation in birds, as well as in a stimulation of drug metabolism in other animals. In the case of birds, we have seen that the increased oxidation of steroids includes the hormone which regulates calcium deposition in eggshells, so that the PCB's may add their influence to the insecticides in producing the thin eggshell syndrome which has threatened a number of species of hawks and other predators.

## Oil Pollution

Like PCB's, pollution by petroleum products is very much in the popular press. The great oil spills have produced catastrophic biological effects and, of course, equally catastrophic effects on esthetic features of the environment. Since throughout much of evolutionary time oil was deep in the earth and thus largely isolated from the biosphere, organisms have generally been rather handicapped in ways to handle it metabolically. For instance, it has required research to find bacteria able to incorporate petroleum fractions into their cellular carbon compounds. This goal, which has met with some success, is important because (1) it seems useful to have a mechanism for converting the hydrocarbons in oil into edible protein and (2) because it would be equally desirable to have a bacterium able to convert spilled oil, say, on the ocean surface, into a harmless material. Our present array of techniques for cleaning up oil spills, which include dispersing detergents, straw, and other absorbants, seem primitive in the extreme, and it would be helpful to enlist the support of the bacterial kingdom.

Petroleum is a complex mixture of hydrocarbons and other carbon compounds many components of which are undoubtedly harmful to cells. For instance, there are certainly carcinogenic compounds (or that can be converted into carcinogens) in oil. However, in general the major effects of oil pollution should probably be regarded as physical effects, rather than as direct effects of particular chemicals on cellular processes. For example, a major effect of oil on the ocean surface is the alteration of that very surface by formation of a thin layer of hydrocarbon on the surface. This thin film, which is highly insoluble in the water below it, is probably much more significant to life in the sea than are any chemical effects of the actual compounds involved. Thus, sea birds are killed because their feathers have become gummed up by the oil with resultant changes in flotation or heat exchange. Likewise, organisms living in the surface layers will be influenced by changes in the surface tension produced by the thin film as well as by altered gas exchange across the ocean surface. The harming of free-swimming organisms like fish may be due to alteration of surface properties of their gills with attendant disruption of oxygen exchange.

In general, huge amounts of oil are required to kill animals when compared to, say, the insecticides, and the destruction appears to be on a level other than cellular. For instance, lobsters, which are very sensitive to insecticides, are only affected by much larger amounts of oil products. Indeed, Maine lobstermen learned long ago that lobsters are actually attracted to petroleum products, probably in the same way that they are attracted to oily fish on which they feed. So great is their attraction that certain unscrupulous fishermen have been known to bait their traps with old crankcase oil, which has no apparent deleterious effect on the lobster save that it ends up in the pot. Unfortunately, the oil ingested by the lobster does not improve flavor and, in fact, such lobster meat tends to taste like the floor of an auto body shop. But such considerations are beyond the scope of the present book!

## Carcinogens and Mutagens in the Environment

Mutagens are compounds that produce mutations; mutations are heritable changes in organisms. Thus, if one observes a population of bacteria (or humans) for awhile, one will be amazed by the high degree to which successive generations resemble each other with, for instance, blond parents producing blond children, and so forth. Occasionally, however, one will observe an abrupt change. For instance, the blond parents may produce a dark-haired child. If this exception is the result of a mutation, the change will be inherited in subsequent generations.

Since human hair color is a complex case, a better example is the mutation that occurs when a population of insects is exposed to insecticides. In this case, perhaps one insect in a million will mutate each generation to a form that is immune to whatever insecticide is being used. The mutation rate is normally independent of the presence of the insecticide, which serves to provide a selective process for spotting the mutant. If the mutation rate is observed to rise on application of the insecticide, we would say that the insecticide is, itself, a mutagen. This is not an impossibility, although unambiguous cases are difficult to find. Finally, the change produced by a mutation is heritable; thus, the offspring of the insecticide-resistant insects will also exhibit the resistance. And, of course, the heritable character of resistance is the bane of those who are in the business of using chemicals to kill insects.

There are two phenomena that may, with some reservations, be regarded as special cases of mutagenesis. First, some chemicals and some physical influences, such as forms of ionizing radiation, cause tumors. Tumors are rather independent collections of cells that grow at the expense of the remainder of an organism. Malignant tumors grow rapidly and invade normal tissue so that they are highly threatening to life; benign tumors are less invasive and are usually not as dangerous. One way of regarding the origin of tumors is that they are the results of a mutation in the cells of the organism—a so-called somatic mutation. The mutation involves a change to a cell form which has properties of uncontrolled growth and invasiveness. In fact, the mechanism by which tumors begin their independent existence is complex and, in general, unknown. It is interesting, however, that many compounds and other influences producing mutations in a variety of organisms also produce tumors, that is, they are carcinogens. Since mankind has an understandable concern with avoiding tumors, there has been considerable furor about the discovery that many components of pollution have either mutagenic or carcinogenic activity. For example, the recent discovery that DDT can give rise to mutations in laboratory mice has been a factor in arguments against its indiscriminate use.

Production of birth defects by chemicals and radiation is also considered closely related to mutagenesis and carcinogenesis. As in the preceding case, one should not be too firm about the extent of similarity. However, some types of birth defects can be regarded as the result of a mutation occurring during early embryonic development, for instance, as a consequence of the mother's ingestion of a harmful drug. Although this is certainly a simplification, it does provide a general basis for grouping the three effects together here.

New compounds are constantly being identified as contributing to mutations, carcinogenesis, birth defects, or all three. A number of drugs have been withdrawn from the market owing to their harmful effects in these respects. Perhaps the most disastrous example is thallidomide, which produced many examples of deformed children until it was eliminated. In other cases the situation is much more subtle. When new compounds are exposed as belonging to these categories (an event that occurs with increasing frequency), the industry producing them often responds that "just about anything is mutagenic" (or carcinogenic or produces birth defects) provided that enough of the compound is used. Since this remark is often justified, one is faced with difficult questions about how much carcinogenesis is acceptable, and people who are worried about tumors tend to have profound difficulty with this kind of analysis.

Furthermore, since experimentation with humans is impossible for moral reasons, one must always base judgments about estimation of mutagenesis, etc., on studies with animals. One is reminded of the debate that raged in the United States when it became apparent that herbicides being used by the armed services in an "undeclared war" had been shown to produce birth defects in laboratory animals. Opinion favoring the use of such compounds argued that it was impossible to draw conclusions about the likely effects on humans from such studies. Although such a sentiment was not reasonable, it might be considered to exhibit a somewhat cavalier attitude toward the health of citizens of Southeast Asia.

In the present discussion, the main conclusion is that a number of compounds, including some components of pollution, have either mutagenic, carcinogenic, or birth-defect-producing effects and that careful surveillance is mandatory. A great amount of further information is required about the mechanisms underlying these effects, and in the meantime mankind would do well to err in the direction of caution.

## Physical Pollution: Radiation

Radiation has been demonstrated to produce birth defects, mutations, and tumors. Various forms of radiation can have these effects with, for example, ultraviolet light being a mutagen often used in the laboratory, as well as causing certain forms of skin tumor. Ionizing radiation, which has sufficient energy to strip electrons from molecules and produce ions, leads to all three

**Table 8-1. Types of radiation.**

| Class of radiation | Type | Nature | Environmental source | Mutagenesis |
|---|---|---|---|---|
| Nonionizing | visible light | electromagnetic | the sun, etc. | 0 |
|  | ultraviolet | electromagnetic | the sun, etc. | + |
|  | microwave | electromagnetic | radio transmission, ovens and other devices | 0 |
| Ionizing | alpha radiation | 2 protons + 2 neutrons | natural and man-made | + |
|  | beta radiation | electrons | natural and man-made | ++ |
|  | gamma radiation | electromagnetic | natural and man-made | +++ |
|  | x-rays | electromagnetic | man-made (eg. dental x-ray and color television) | +++ |

Note: Ionizing radiation is defined as its ability to form ions (i.e., remove electrons) from molecules in solution, a property that reflects its higher energy. The reason that gamma radiation and the similar x-rays may be regarded as more mutagenic than the other forms is their considerable ability in penetrating matter.

categories of harm, and forms of it such as gamma radiation can penetrate deeply into organisms, extending the location that can be influenced.

Because the effects of radiation on cells represent a field of biology noted for its complexity, we shall provide only a few comments. First, the mutagenic effects of radiation on cells may, in the most simple case, be thought of as direct damage to the molecule that "contains" the genetic information, the DNA. Thus, radiation may simply cause breaks in the DNA chain, and these will produce mutations as well as, in some instances, spectacular alterations in the chromosomes, visible with an ordinary light microscope. Such chromosomal aberrations are associated with many different birth defects, including forms of mental retardation. They also appear to be produced by chemical agents as well as radiation. The ability of radiation to produce such effects obviously builds a strong case against indiscriminate exposure of pregnant women to x-rays.

Besides damaging DNA, radiation harms cells by direct "hits" on enzyme systems, and some systems are much more susceptible than others. Finally, some radiation effects on cells occur due to the interaction of the radiation with water and solutes within the cell, leading to the production of peroxides and various free radicals which are themselves harmful in various ways. Free radicals are compounds which have had an electron removed and which are, therefore, highly reactive. Peroxides are also highly reactive, having a structure R—O—O—H that has the relatively unstable oxygen-oxygen configuration. Peroxides are thus sufficiently toxic to cells that most cells have a special enzyme called catalase which breaks them down to harmless components such as oxygen and water. Because peroxides themselves are mutagenic, radiation could be said to get at least two damaging attacks on the DNA and consequent production of heritable damage.

## Physical Pollution: Heat

Another form of physical pollution is the thermal pollution that is of major interest to people who design (and also who try to prevent the designing of) power plants. Cells can live and prosper only within a restricted range of temperatures, and the heating of water occurring when industries use environmental water as a coolant often reaches a level that can kill, or otherwise discourage, many organisms. Two aspects of heat damage to cells bear mention in this context. First, heat is able to destroy the three-dimensional structure of proteins. This process, called *denaturation,* occurs through the increase in thermal motion that irreversibly alters the protein structure. Since enzymes are always proteins and are essential for the function of cells (see Chapter 3), sufficient heat to denature protein is also sufficient to kill the cell. Most cells are rapidly killed when the temperature rises to 50° or 60°C., although there is considerable variation among different organisms.

Thus, a borderline level of thermal pollution will discourage some cells but not others, changing the balance of species in the water. Since increasing temperatures lead to increasing rates of growth for many organisms until denaturing temperatures are reached, the same increase that kills some organisms can promote growth of less sensitive ones, accelerating changes in the population composition. Finally, an extreme example of variation in heat sensitivity is seen in organisms, usually blue-green algae and bacteria, that live in very hot waters such as hot springs. These cells obviously survive in waters hot enough to kill most cells, and it is most significant that their proteins are also not denatured until much higher than normal temperatures are reached.

The notion that cells can be killed by immersion in boiling (or nearly boiling) water is a rather unsubtle one. In fact, cells and thus organisms can also be harmed indirectly by existing at lower but still excessive temperatures which do not lead to protein denaturation. Rates of chemical reactions are increased by being allowed to proceed at elevated temperatures. The reactions comprising cell metabolism are such that their rates double when the temperature is raised roughly 10°C. This means that a higher than normal temperature has the effect of stimulating metabolic reactions in organisms which must exist at the same temperature as their surroundings, i.e., those organisms that are not warm-blooded. The important set of reactions of cellular respiration—the reactions leading to oxygen consumption—are increased in rate when organisms are warmed up. Fish and other aquatic organisms respire more rapidly and thus consume more oxygen at high temperatures. This increase is especially serious because the same elevated temperature lowers the amount of oxygen that can dissolve in water. Increased temperature leads to a greater need for oxygen and, at the same time, a lower oxygen content to satisfy that need. The two effects converge to make life unsupportable for many organisms when water is heated artificially because of a power plant or other source of hot water.

# Chapter 9

# Conclusions and Suggestions for Further Reading

The main conclusions of this book are that (1) the toxic effects of pollution on cells are complex but that (2) it is exceedingly important to be informed of them. This book is a primer in that it attempts to introduce the subject of pollution physiology on a level accessible to beginning biology (and other) students, and the requirements for such a book dictate that the book actually discuss a fraction of what is known. Moreover, it is clear that such a subject is in its early infancy; it is hoped and expected that the next five years will see a vast increase and improvement in the information available. If this is the case, the reader may wish to keep abreast of new developments; since the popular press is probably not an adequate source of this type of technical information, the reader may welcome the list of readings and additional comments that follow. These comments and readings should be considered as a portion of a strategy for becoming more informed about this field, and not, in any sense, as a definitive list.

Ignorance seldom provides an effective basis for accomplishment. In the area of pollution abatement, accomplishment is the requirement, and this book proposes that detailed information about all aspects of pollution as it impinges on the biological world is likely to produce practical and unexpected benefits. Students, therefore, are urged to become as widely informed as possible about the fundamental science required to understand the properties and effects of substances comprising pollution. A disconcerting feature of the modern age and of the realities involved in improving mankind's conditions is that a pure heart is not enough; information is usually required and the acquisition of such information is sometimes unsatisfying and unforgiving, but quite necessary.

We appear to be at the threshold of a new understanding of the relation between man and the remainder of the natural world. One element of that understanding is the realization that man has altered the world in a fundamental and irreversible way. That alteration has occurred through ignorance and arrogance but it is, nonetheless, a fact, and the developing ecology that is being embraced is novel, largely in recognizing the fundamental force produced by the industrialization of human society.

We appear to be engaged in combating man's ignorance about his world and it is even possible that some erosion of the arrogance may occur, but the problems will remain with us. The material that we have dumped into the sea will persist there for, in some instances, a period comparable with the geological time-scale. Meanwhile, these fragile and wonderful objects —the cells of living organisms—must adapt to the new situation; existing signs are that they will somehow survive and prosper. But they are truly fragile (as we are fragile in the final sense), and we may not exceed the limits of tolerance without harm. Obviously it will be of central importance to discover what those limits really are and what features of cells serve to place boundaries on them. This sort of study has only begun and not a moment too soon.

## Suggestions for Further Reading

These suggestions represent only a few of the many books available in the various categories. The selection reflects prejudices of the author as well as a desire to avoid overwhelming the reader with the mass of recent publications in this area. Books in each section are listed in order of publication date.

### General Works About Pollution

Rachel Carson, *Silent Spring*, 1962, Houghton Mifflin Co., New York (Available in paperback).

Robert Rudd, *Pesticides and the Living Landscape*, 1964, The University of Wisconsin Press, Madison (A paperback).

The American Chemical Society, Subcommittee on Environmental Improvement, Committee on Chemistry and Public Affairs, "Cleaning our Environment—The Chemical Basis for Action," 1969, American Chemical Society, Washington, D. C. (A paperback).

John Harte and Robert Socolow, *Patient Earth*, 1971, Holt, Rinehart, and Winston, New York (A paperback).

Charles Warren, *Biology and Water Pollution Control*, 1971, W. B. Saunders Company, Philadelphia.

H. Stephen Stoker and Spencer Seager, *Environmental Chemistry*, 1972, Scott, Foresman and Company, Glenview, Illinois (A paperback).

Kenneth Wagner, Paul Bailey and Glenn Campbell, *Under Seige*, 1973, Intext Educational Publishers, New York.

## General Ecology

Edward Kormondy, *Concepts of Ecology*, 1969, Prentice-Hall, Inc., Englewood Cliffs, N. J. (A paperback).

Eugene Odum, *Fundamentals of Ecology*, 3rd Edition, 1971, W. B. Saunders Company, Philadelphia.

Robert Ricklefs, *Ecology*, 1973, Chiron Press, Newton, Mass.

## Cell Physiology and Biochemistry

Albert Leninger, *Biochemistry*, 1970, Worth Publishing Co., New York.

Robert Dowben, *Cell Biology*, 1971, Harper and Row, New York.

Stewart Edelstein, *Introductory Biochemistry*, 1973, Holden-Day, Inc., San Francisco.

John Howland, *Cell Physiology*, 1973, Macmillan, New York.

## Pesticides

R. D. O'Brien, *Insecticides—Action and Metabolism*, 1967, Academic Press, New York.

Morton Miller and George Berg, *Chemical Fallout*, 1969, Charles C. Thomas, Springfield, Illinois.

R. D. O'Brien and Izuru Yamamoto, *Biochemical Toxicity of Insecticides*, 1970, Academic Press, New York.

## Air Pollution

Richard Scorer, *Air Pollution*, 1968, Pergamon Press, Oxford (A paperback).

Arthur Stern, *Air Pollution*, 2nd Edition, 1968, Academic Press, New York.

Samuel Butcher and Robert Charlson, *An Introduction to Air Chemistry*, 1972, Academic Press, New York.

## Metal Pollution

E. J. Underwood, *Trace Elements in Human and Animal Nutrition*, 1972, Academic Press, New York.

Douglas Lee (editor), *Metallic Contaminants and Human Health*, 1972, Academic Press, New York.

The serious reader will undoubtedly want to go beyond the level of treatment in any of these books, and scientific books are, in general, rapidly outdated anyway. Review articles provide the next level of information, being no more than a year out of date when they appear. Important review articles have appeared in a wide variety of sources such as the Annual Reviews of Pharmacology and Biochemistry. A particularly helpful review of pesticide biochemistry by J. E. Casida appeared in Annual Review of Biochemistry, Volume 42, page 259 (1973).

Other review articles are published frequently in symposia proceedings devoted to environmental matters. Such symposia are being held with increasing frequency. Finally, the serious student of matters related to pollution will necessarily have recourse to primary journal articles and to the several abstract services existing in this pivotal area. There are journals devoted to the field, including the *Journal of Environmental Contamination and Toxicology*. Many papers of interest appear in journals devoted primarily to other disciplines, such as physiology, ecology, or cell biology. Due to the interdisciplinary character of the field, as well as to the frequency with which new journals and abstracting services appear, it is not desirable to enumerate such sources; rather, it is important to note that the books mentioned above, as well as review articles, will provide in their lists of references a valuable guide in locating further information.

# Appendix

## Basic Organic Structures

A reader who has not studied organic chemistry should find the following comments helpful in understanding the structures which appear in various portions of this book. First, organic chemistry is almost synonymous with the chemistry of carbon and does not necessarily imply that particular organic compounds have relation to the processes of life (although, of course, many of them do). In fact, carbon, owing to its specific bonding properties, does play a central role in the chemistry of life, partly because of its ability to form strong bonds with other carbon atoms, leading to the possibility of forming long carbon chains. These serve as a backbone for many of the large molecules of living processes, and they provide for the richness of chemical biology. Carbon is able to form bonds with many other atoms, thus allowing for the variety of compounds which appear on the pages of this book and in the living world. But in most instances, the predominant atom is carbon itself.

### Bonds

Bonds, which are denoted in a two-dimensional structure as a dash as in (C—C), reflect attractive forces that bind two atoms together. The strength of these bonds, as well as their geometry, is specific for the particular atoms involved. For instance, the energy required to break a carbon-carbon bond (and therefore its stability) is approximately twice that of a nitrogen-nitrogen bond, and the interatomic distances are somewhat different. The energy of bonds, as well as the number of bonds that can be formed by a particular atom, may be deduced from a quantum-physical description of the atomic structures involved—a topic far removed from the scope of this book. Our purposes will be satisfied by a modest description of molecular structures in terms of some simple substructures called *chemical groups*.

For example, the number of bond-forming electrons (i.e., outer or valence electrons) determines the number of bonds in which a given atom can participate. This is because most bonds in organic molecules reflect the sharing of such electrons between atoms. Thus, carbon is generally able to form four bonds; hydrogen, one; oxygen, two; nitrogen, three; and so forth. The arrangement of bonds around an atom is often such that the bonds do not lie in a single plane. Thus, carbon exhibits a tetrahedryl geometry with the atom in the center and the bonds being "located" at the points of the tetrahedron. When we draw structures on the page, we ignore this three-dimensional character and draw carbon, for instance, in the center of four right-angle bonds as seen in the following structures.

## Double Bonds

Oxygen can form two bonds as in the case of the water molecule,

$$H-O-H$$

where they attach two separate hydrogens. On the other hand, sometimes the two bonds go to the same atom as in the carbonyl group,

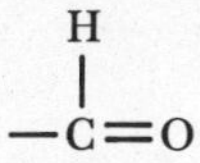

$$\begin{array}{c} H \\ | \\ -C=O \end{array}$$

and this is called a *double bond*. Such bonds are widespread in organic chemistry, occurring commonly with carbon, nitrogen, and phosphorous compounds. Indeed, in the case of nitrogen and carbon, triple bonds occur as well, an appropriate example being hydrogen cyanide.

$$H-C\equiv N$$

Double bonds are more reactive than single ones with triple bonds being more active still.

## Chemical Groups

The special bonding properties of the various atoms lead to the common occurrence of certain characteristic clusters of atoms in a wide variety of compounds. These clusters—chemical groups—may be regarded as the building blocks of compounds. The groups give many compounds their characteristic chemical and physical properties as, for example, when the acidic carboxyl group

$$-COOH$$

renders a compound an acid. The groups are also the basis for systematic chemical names for compounds so that knowing the more common ones greatly simplifies chemical nomenclature.

Table A-1 identifies some of the more common chemical groups as well as characteristic properties. The latter column outlines the contribution of a particular group to the properties of compounds that contain it and may also serve as an aid to memory.

Table A-1. Some common chemical groups.

| Class | Name of group | Structure | Characteristic |
|---|---|---|---|
| 1. Contain C and H only (alkyl groups) | methyl | $CH_3-$ | All are relatively unreactive. |
|  | ethyl | $CH_3CH_2-$ |  |
|  | propyl | $CH_3CH_2CH_2-$ |  |
| 2. Oxygen-containing | carboxyl | $-\overset{O}{\underset{}{C}}-OH$  $-COOH$ | or acidic |
|  | carbonyl | $-C=O$ | Makes a compound an aldehyde or ketone. |
|  | hydroxyl | $-OH$ | Found in all alcohols. |
| 3. Nitrogen-containing | amino | $-NH_2$ | Found in amino acids. |
|  | nitro | $-NO_2$ |  |
| 4. Other | phosphate | $-OPO_3H$ | Renders compound weak acids. |
|  | sulfhydryl (or mercapto) | $-SH$ | Binds heavy metals. |
|  | acetate | $CH_3\overset{O}{\underset{\|\|}{C}}O-$ | Acetic acid is this structure with an additional H at the bond. |
|  | phenyl | $\langle\bigcirc\rangle-$ | Fairly unreactive but makes compound more soluble in organic solvents. |

## Ring Structures

Many compounds contain rings of atoms, a number of examples occurring among the insecticides. These rings may contain only one kind of atom as in the case of benzene,

or they may contain more than one as in the case of pyridine.

Compounds also contain rings that are fused to other rings, an excellent example being the amino acid, tryptophan.

Many rings are connected by ordinary bonds identical to those that connect nonring compounds. For instance, the properties of cyclohexane

are exceedingly similar to those of the corresponding straight-chain hexane.

$$CH_3-CH_2-CH_2-CH_2-CH_2-CH_3$$

Conversely, there is a category of ring compounds called *aromatic compounds* which have quite characteristic properties unlike those of the corresponding straight-chain compound. Aromatic compounds are often written as possessing alternating single and double bonds (e.g., benzene)

## 120    Appendix

```
      H
      C
   HC   CH
   HC   CH
      C
      H
```

or they are written with single bonds and a circle in the center.[1]

In either case, the bonds should be regarded as hybrid, including a regular component "between" the atoms as well as a component that is delocalized throughout the entire compound.[2] In any case, such compounds exhibit chemical properties different from ordinary compounds and are, in general, more stable than would otherwise be the case. With regard to nomenclature, when the benzene ring appears as a group in a larger compound it is called a *phenyl group*.

Finally, compounds can be based upon quite complicated ring structures. For instance, a structure that has been mentioned several times in this book is heme

```
         C   N   C
       N ···· Fe ···· N
         C   N   C
```

which has, as its basis, four five-membered rings, each containing four carbons and one nitrogen. Heme underlies several important molecules in cells including the oxygen-carrying blood pigment, hemoglobin, the cell respiration compounds, the cytochromes, and the major photosynthetic pigment, chlorophyll, where the iron is replaced with magnesium.

---

1. By convention, individual carbon and hydrogen atoms are often not written in the ring structure.
2. In chemical terms, some valence electrons are shared between the atoms while the rest are delocalized in a so-called pi cloud of electrons, parallel to and above and below the plane of the ring of atoms.

# Index

acetylcholine esterase 50, 71
active transport 46
adenosine triphosphate (ATP) 41, 46; ATP-ase 46, 47, 48; ATP synthesis 24, 25, 42
aldrin, effect on fish 82
amino acids 28, 29
antibody 54
aromatic compounds 119
autotrophic organisms 8

benzoic acid 55, 56
beryllium 103
binding sites 54
biodegradability 17
biological cycles 15
biological membrane 33

cadmium 102
carbamates 67
carbon cycle 16
carbon dioxide 84
carbon monoxide 84, 85, 87, 88; effect on nitrogen fixation 89; physiological effects 86
carcinogens, environmental 107
cell structure 19, 20
chemical bonds 116
chemical groups 117
chlordane 59, 66, 67; effect on fish 82; nontarget effects 78
chloroplasts 24, 25
cholesterol 31
choline 71

chromatin 22
chromium 103
cyclodienes 67
cytochromes 41, 42
cytochrome oxidase, effect of carbon monoxide 88
cytoplasm 19

DDT 13, 53; action on nervous system 69, 70; as mutagen 107; effect on fish 82; effect on membranes 83; intoxication 69; nontarget effects 78; stimulation of microsomal oxidases 58, 61; toxicity 66
denaturation 110
detoxification mechanisms 55, 56, 57
dieldrin, toxicity 66
DNA 22, 43, 44
double bonds 117

ecosystem 6, 7, 9
endoplasmic reticulum 20, 26, 57, 59
environmental toxicology 3
enzymes 35; active site 36; inhibition 37
estradiol 80
ethylene 94

fermentations 40
fixation of carbon dioxide 15
flavoprotein 73, 75

fluorine 95
food chains 8, 11

golgi complex 26

heat pollution 110
hemoglobin 87, 89, 102
hippuric acid 56
hydroxylation, in detoxification 57

insecticides 1, 10, 73; effect on birds 78; naturally occurring 73; site of attack 50, 68
integument 52
ionizing radiation 108

$LD_{50}$ 64, 65
lead 53, 100, 101; cellular action 102; organic forms 101; poisoning 100
lobster, effect of DDT 82
lysosomes 27

marine algae, insecticide sensitivity 83
membrane barrier 51
membranes 31, 45
mercury 53, 54, 96, 97, 98, 99; effect on proteins 99; organic compounds 98, 99
messenger RNA (mRNA) 43, 44, 45
microsomal oxidase 57, 58
microsomal oxidase system 59, 60, 61
mitochondria 20, 22, 23, 24
mixing time 12
mutagens, environmental 107

NADH oxidation, effect of rotenone 74
nerve cells 47, 49
nicotine 76; structure 76
nitric oxide 90
nitrogen dioxide 90
nitrogen fixation 89
nitrogen oxides, influence on the lungs 93

nitrous oxide 91, 92
nuclear envelope 20
nucleolus 20, 22
nucleus 19, 20, 21, 22
nutrient cycles 15

organic peroxides 92, 93
organic phosphates 67; action of 70, 71
organic structures 116

parathion 67, 72
peptide bond 29
peroxides, produced by radiation 110
persistence, of insecticides 66, 68
pesticides, resistance to 14
phenylacetic acid 56
phenylmercuric acetate 98
phenylsulfuric acid 57
phospholipid 31, 33
phospholipid bilayer 34
photosynthesis 16, 19, 24, 25, 40
pica 100
piperonyl butoxide, as a pesticide synergist 61
plasma membrane 19, 20, 53
pollution, atmospheric 84
pollution, effects of 13, 14
pollution, metal 95
pollution, oil 105, 106
polychlorinated biphenyls 104, 105
primary consumers 8
primary production 8
protein 22, 28, 30
protein synthesis 43
pyrethrin 67; toxicity of 66
pyrethrins, action of 76

radiation 109, 110
residence time 11
respiration 16, 21, 24, 39
respiratory ATP synthesis 42
respiratory chain 41, 42; effect of rotenone 73
resting potential 47
ribosomes 20, 26, 27

ring structures   119
rotenone   54, 67, 73;   action of 73;   structure   75

salmon, effect of DDT   82
secondary consumers   8
serine, in acetylcholine esterase   72
sevin   67
skin, human   52
smog   85
sodium-potassium pump (ATPase)   46, 47, 48;   inhibition by DDT   83
solubility   9, 10
steroid   58

synapse   49, 50
synergists   61, 65

thin eggshell phenomenon   79
threshold concentration   4
thylakoid membranes   25
toxicity, measure of   64
trace elements   95
transfer RNA (tRNA)   44, 45

vacuole   20
vanadium   103
vitamin D   80

zinc   102;   inhibition of cytochrome system   103